TIMBER

The EC Woodcare Project: Studies of the
behaviour, interrelationships and management
of deathwatch beetles in historic buildings

ENGLISH HERITAGE

ENGLISH HERITAGE RESEARCH TRANSACTIONS
RESEARCH AND CASE STUDIES IN ARCHITECTURAL CONSERVATION

TIMBER

The EC Woodcare Project: Studies of the
behaviour, interrelationships and management
of deathwatch beetles in historic buildings

EDITED BY

Brian Ridout

Volume **4**

July 2001

JAMES X JAMES

First published by James & James (Science Publishers) Ltd,
35–37 William Road, London NW1 3ER, UK

A catalogue record for this book is available from the British Library
ISBN 1-873936-65-6
ISSN 1461 8613

Volume editor: Brian Ridout
Series editor: David Mason, English Heritage
Consultant editor: Kate Macdonald

Printed in the UK by Hobbs The Printers

Disclaimer
Unless otherwise stated, the conservation treatments and repair methodologies
reported in this volume are not intended as specifications for remedial work.
English Heritage, its agents and publisher cannot be held responsible for any
misuse or misapplication of information contained in this publication.
The inclusion of the name of any company, group or individual, or of any
product or service in this publication should not be regarded as either a recom-
mendation or endorsement by English Heritage or its agents.

Accuracy of information
While every effort has been made to ensure faithful reproduction of the original
or amended text from authors in this volume, English Heritage and the publisher
accept no responsibility for the accuracy of the data produced in or omitted from
this publication.

Front cover
Adult *Xestobium rufovillosum* (deathwatch beetle)
(photo: Brian Ridout).

Contents

Acknowledgements

English Heritage and its editors hereby acknowledge the European Commission's Directorate-General for Science, Research and Development [formerly DG XII] for financial support towards the underlying costs of the *Woodcare* research project upon which findings an international conference and this volume of the Research Transactions have been based. The Commission's research contracts manager, Dr Julia Acevedo Bueno, greatly assisted in expediting and steering the project. The costs of the *Woodcare* conference and of these proceedings, however, were entirely borne by English Heritage as part of the organisation's continuing commitment to sharing and promulgating the results of technical and scientific research in architectural conservation.

The *Woodcare* research was undertaken by an international consortium of four partners in three countries of the European Union. Representing the United Kingdom, project co-ordination was undertaken by English Heritage and led by John Fidler with technical and administrative support from Chris Wood, Jo Rodgers and Amanda Holgate. Scientific co-ordination was led by Dr Brian Ridout of Ridout Associates under a subcontract to English Heritage. Collaborating English partners included Birkbeck College London, led by Dr Wally Blaney, now Emeritus Professor of Biology. Steven Belmain was Professor Blaney's assistant: then a doctoral student and now holder of a PhD working for the Natural Resources Institute, University of Greenwich. Additional experimental work for Birkbeck was carried out by Professor Monique Simmons of the Jodrell Laboratory, Royal Botanic Gardens, Kew.

The Dutch team involved staff from a variety of departments at TNO-Bouw, the national building research establishment in Delft. Leadership was provided by Dr Petra Esser of the Department of Wood Science with scientific and technical support from Dr P van Staalduinen, Dr A Tas and Jan de Jong. The Irish team from University College Dublin consisted of Dr Dervilla Donnelly, Professor of Chemistry and Dr Hubert Fuller of the Department of Biology, with the assistance of Colm Moore, a doctoral student.

The *Woodcare* conference programme was devised by Dr Brian Ridout in association with English Heritage. Dr Ridout and staff from Ridout Associates helped to administer the conference along with Chris Wood and Amanda Holgate from English Heritage.

English Heritage is grateful for the excellent presentations of ancillary complementary research delivered at the conference by Drs Wyatt and Birch of Oxford University, by Dr Graham Coleman and Robert Demaus. Their papers are included in the following chapters. Additional supporting papers on structural timber repair techniques that provided a field context for the *Woodcare* research were also presented by Graham Pledger of English Heritage's Conservation (Structural) Engineering team and by master carpenter Peter McCurdy but are not included here.

English Heritage acknowledges the help and assistance of Paul Perry and his team from the Salvation Army Citadel, London, for the provision of conference facilities at short notice.

The organisation is also grateful for the technical co-operation of the British Wood Preservation and Damp-proofing Association (BWPDA) and the financial support of conference exhibitors and sponsors without whose generous patronage the meeting would have been less successful.

Brian Ridout acted as the guest scientific editor for this volume of the Research Transactions. Dr Kate Macdonald was the publication's project manager and consultant editor and Dr David Mason was the series editor.

Preface

JOHN FIDLER

The papers contained in this volume of English Heritage's Research Transactions are a formal record of the *Woodcare* conference, held in London on 23 and 24 September 1998. This international gathering concerned itself with technical and scientific questions related to structural oak timber decay in historic buildings and studies of deathwatch beetle (*Xestobium rufovillosum*), the virulent pest species found in north-west Europe but also extant in similar temperate zones across the world.

Eleven of the sixteen papers delivered specifically reported on the findings of the European Commission's *Woodcare* project: contract EV5V CT940517 (1994/5–1997/8) and fulfilled a binding contractual commitment to deliver scientific results to practitioners. Funded as an international collaborative scientific research effort as part of the European Union's Fourth Framework *Environment R&D* programme, the work was supervised by what is now the Commission's Directorate-General for Science, Research and Development (formerly DG XII). The research sought to study the interaction of deathwatch beetles and timber in historic roof spaces, taking into consideration the ages of timber; the relationship of beetles to fungi (in particular *Donkioporia expansa*, the oak rot fungus), natural predation and the environment.

Additional scientific and technical papers presented at the conference, notably those by Drs Wyatt and Birch on the mating behaviour of the deathwatch beetle, by Dr Graham Coleman on pesticides and by Robert Demaus on nondestructive diagnostics as tools for structural timber infestation assessment, directly complement and reinforce the *Woodcare* work and are reported here.

The meeting brought together architects, surveyors, engineers, carpenters, timber treatment specialists, pesticide chemists, wood scientists, biologists, mycologists and entomologists from the United Kingdom, The Netherlands and the Irish Republic. The papers demonstrated how international practice in the remedial treatment business for the repair and preservation of historic structural timber has changed over the course of the twentieth century to become more subtle and conservative.

As a result of the research undertaken here, less interventive refurbishment processes, brought about by a better understanding of pest behaviour in the field, can now reduce the building industry's reliance on chemical palliatives. An integrated approach to pest management, which understands the fallibilities of building construction and maintenance, appreciates the inefficiencies in preservative and pesticidal treatments and instead utilises environmental controls, encourages natural predation and focuses on the inter-relationships of the deathwatch beetle with its habitat, can now lead to a cleaner, more benign and cost-effective control of infestations for the benefit of building occupants, construction operatives and the built heritage.

The *Woodcare* project's entomological research, explained in the following pages, and additional mycological work carried out by the editor of this volume, Dr Brian Ridout, has enabled English Heritage and its Scottish counterpart, Historic Scotland, to amend their technical policies towards structural timber decay and its treatment by adopting a conservative approach (Ridout 2000).

I commend these papers to readers. Applied science in the service of architectural conservation has rarely been better directed.

John Fidler RIBA
Head of Building Conservation & Research,
English Heritage,
Coordinator of the EC *Woodcare* Project EV5V
CT940517 (1994/5–1997/8)

Ridout B, 2000 *Timber Decay in Buildings and its Treatment: A Conservation Approach*, English Heritage and Historic Scotland with E & F N Spon Ltd, London.

An introduction to the Woodcare project

JOHN FIDLER

English Heritage, Building Conservation Research Team, 23 Savile Row, London W1S 2ET, UK;
Tel: +44 (0)20 7973 3025; Fax: +44 (0)20 7973 3130/3249/3001; email: john.fidler@english-heritage.org.uk

Abstract

This paper briefly describes English Heritage's strategic technical research programme in its cathedral grants scheme that provided the initial assessment of structural timber infestation and decay caused by deathwatch beetle and made recommendations for future scientific research. This led to collaborative involvement in the European Commission's Woodcare research project whose genesis is subsequently described and explained.

Key words

English Heritage, technical research, European Commission, the Woodcare project, deathwatch beetles, oak rot, timber decay, timber ageing, cathedral roof timbers and their environment.

BACKGROUND: ENGLISH HERITAGE RESEARCH

Between 1991–99, the annual budget for scientific research on building materials within English Heritage averaged £200–£250,000 per year, rising to almost £600,000 at peak times. These sums, constituting less than 2% of the organisation's total budget, enabled the organisation's Building Conservation and Research Team (BCRT) to work on a variety of technical studies of decay in historic building materials and their treatment, and to publish the results.

BCRT's twenty five projects range from engineering experiments concerned with historic mortars and their performance, through studies of the decay processes in and remediation of deteriorating sheet metal roof coverings, to investigations of masonry consolidants, stained glass conservation and timber decay and treatment. Over the past ten years, thirteen sets of contractors have been commissioned, employing more than thirty five specialists in different scientific disciplines: all of them run from an interdisciplinary office of architects, conservators and surveyors expert in architectural conservation (Fidler 1996).

From its inception in April 1984, English Heritage[1] always sponsored research. Its predecessor organisations, the Ministry of Works and, later, the Department of the Environment's Directorate of Ancient Monuments and Historic Buildings (DAMHB), both had modest budgets for experimenting with old materials to see how they worked and how they could be repaired. But in 1991 the then Prime Minister, Margaret Thatcher, became concerned about the growing burden of decay facing England's major cathedrals and awarded English Heritage significant additional once-only funds to invest in these major listed buildings: to understand their deterioration and facilitate their repair (Fidler 1998).

English Heritage first set about assessing what the precise technical problems were that the cathedrals commonly faced. A summary technical survey of the sixty historic buildings was commissioned from the eminent cathedral architect, Harry Fairhurst FRIBA, who also talked to the cathedral administrators and their architects, to come to some strategic recommendations for concerted action. Out of this work, a programme of grant aid for repairs was developed which continues to run today.[2] A proportion of this special grants budget was then devoted to technical research for the cathedrals, based on the logic of trying to find common answers to collective technical problems. The research was designed in such a way that its outputs could also have an additional impact on understanding the deterioration of other ecclesiastical and secular listed buildings.[3]

Armed with the *Fairhurst Report* (1991) English Heritage initially devised a three-, then five- and ultimately a seven-year phased programme of high priority research projects, one of which involved a preliminary assessment of timber decay in the large medieval oak timber roof spaces that are common to many cathedrals.

At Winchester Cathedral, for example, where the well-recorded and highly documented timber structural system in the roof is said to posses jointing patterns with historical precedence that can be dated back to eighth-century Istanbul (Hewitt 1980), the ingress of rainwater from leaking lead-lined gutters was discovered to have encouraged significant outbreaks of decay, caused apparently by the inter-relationship of fungi and wood-eating beetle larvae. At Salisbury Cathedral a huge infestation of oak rot along wall plates appeared to have gone unnoticed and seemed also to be attracting deathwatch beetles (*Xestobium rufovillosum*).

Dr Brian Ridout of Ridout Associates was commissioned to undertake a representative sample survey of cathedral timber roof structures and their problems of bio-deterioration. His initial findings reported on the rather poor state of general maintenance of cathedral roofs considering their status and that of the buildings. Many rainwater leaks were evident which had apparently led to significant fungal outbreaks of dry rot (*Serpula lacrymans*) and oak rot (*Donkioporia expansa*) with associated beetle infestations.

Ridout's report also revealed a wide variety of technical strategies for timber decay treatments among the many surveyors of the fabric of cathedrals. In some cases, the treatments were extremely drastic, physically eliminating significant amounts of viable historic fabric in 'scorched earth' policies that paid little attention to actual sources and centres of decay. Practitioners appeared fearful for their liability insurance and took excessive measures in reaction to the actual size, scope and intensity of timber deterioration involved. Other treatments were entirely ineffective, either poorly targeted or actually treating moribund infestations in some cases. Dr Ridout recommended additional further study into the actual biological, environmental and ecological issues pertaining to these systems, since traditional remedial treatment systems in general practice did not seem to be working.

One particular set of issues raised related to our old friend, the infamous deathwatch beetle. Although there was a reasonable body of information in the public domain about the beetles, many of the scientific studies pertaining to them were long out of date and few were based on true field studies. It was therefore thought that some of the treatment systems being employed, such as smoke bomb emissions, spray treatments and the like, were not actually addressing many of the infestation problems confronting historic cathedrals and other secular listed buildings.

To assess the significance of Dr Ridout's preliminary findings and to evaluate his scientific recommendations, English Heritage set up a Cathedrals Research Client Liaison Committee[4] that reported to its Cathedrals and Churches Advisory Committee (now called Places of Worship Advisory Committee). This body deemed it essential to invest some of English Heritage's resources in a research programme to study the deathwatch beetle in its environment in roof spaces, particularly to understand the various inter-relationships and interactions.

What became clear at the start of the exercise was the terrible reputation of deathwatch beetle and dry rot fungi among building management bodies in England, generating fear and loathing in equal measures. The general perception seemed to be that one or the other of these infestations were insurmountable and would inevitably lead to very expensive remedial work involving wholesale replacement of infected parts. A great deal of mythology had grown up around the beetles, perhaps out of proportion to the size of the problem involved. But it was also plain that some specifiers and designers charged with the welfare of cathedrals were seriously concerned about their abilities to control them.

As the general understanding of deathwatch beetles and their behaviour was rather rudimentary at the time, other specifiers seemed to panic when confronted with the problem, even when, in reality, they did not have one. At one cathedral, for example, Dr Ridout found long dead beetle outbreaks being regularly treated every spring. Although most conservation architects, surveyors and engineers memorise the life cycle of the deathwatch beetle from textbooks that have been around for years, few actually know from personal detailed observations how the insects behave. Thus there was a real need to revisit the field, which is what the Woodcare research attempted to do.

TRENDS IN THE REMEDIAL TREATMENT INDUSTRY

Ridout's survey confirmed that, once recognition of the timber decay phenomenon had been achieved, the surveyors of the fabric understandably passed the task of eradication to the specialist remedial treatment industry. But few architects or treatment companies appeared equipped to deal with the systematic targeting of outbreaks. More usually, general visual inspections led to the 'carpet bombing' approach to eradication, where whole roof spaces were sprayed or coated with insecticides in order to gain treatment guarantees, thought by all to be an essential output for their clients' peace of mind.

English Heritage and its consultant also noted that the kind of surveying and investigative equipment commonly used for the assessment of structural timber in the cathedrals, and in historic buildings in general, was rather unsophisticated. There have been many technical and scientific developments over the last fifteen years, which have led to more cost-effective ways to evaluate the condition of roof spaces and their construction. So it was thought prudent to review these issues and observe where technical trends were going, to devise, if possible, new and improved technologies to help to locate and target pest infestations and to validate the effectiveness of treatments.

Most of us would agree with another finding from Ridout's report: that there is a widespread growing public concern about the chemical treatment of infestations in buildings. One has only to read the letters pages of coffee-table magazines on home improvements to discover contributions on the 'dangerous' nature of chemicals in the environment. Despite our own experience as professionals in the building industry dealing with sometimes potentially hazardous materials in safe and controlled ways, we must understand lay perceptions of the subject and these genuine fears. Indeed, for a safe and sustainable planet, we must all strive for more benign, cost-effective treatments that are safe for manufacturers, operatives, building occupants and the wider environment. The 'greening' of the timber treatment industry started years ago and we all have a role to play in its long-term success.

I would not like anyone to get the impression from this conference that English Heritage in any way sees *complete* alternatives to chemical insecticidal treatments. Far from it. We would always state that chemical insecticides have a distinct role to play in the eradication of deathwatch beetle infestations.

But we believe, following our research, that there may be more subtle ways that they can be employed. It may soon be possible to allay some, if not all, public concern about the amounts of chemicals necessary for the effective treatment of timbers in roof spaces. The amounts can be reduced by more effective targeting and through the employment of ancillary trapping systems and natural predation. Combined with better control and management of the microclimate and environment (ie keeping water at bay), we may soon be able to devise integrated management systems of benefit to the wider environment, minimising even slight health risks to operatives and occupiers.

CONSERVATION ETHICS

Now, of course, English Heritage comes to this set of problems from a particular perspective. We are a building conservation body governed by a code of ethics encompassed in the tenets of the Venice Charter (ICOMOS 1966, but see also Brereton 1991, DoE/DNH 1994). We speak of activities such as surveying, recording, cleaning, maintenance, treatment and repair as *interventions* in the long lives of historic buildings. We acknowledge that caring for monuments, if not done well, can sometimes be as damaging to them as significant alterations or demolition. So we aim in all our work to make our interventions:

- urgently necessary
- minimal
- if possible reversible
- non-prejudicial to future interventions.

Armed with these guiding principles, we approach the remedial treatment business for timberwork in historic buildings in a particular way. We are concerned that chemical treatments, while being effective for the treatment of the timber itself, are not prejudicial to the welfare of adjoining masonry or decorative plaster, fabric or painted ceilings and the like. So often we have seen brickwork irrevocably damaged by serried ranks of large-diameter drill holes for irrigating the masonry. Salts from the chemical treatments can subfloresce in adjacent masonry or plasterwork, crystallising and causing the materials to exfoliate, powder and crumble into dust. If the chemical treatments themselves do not generate salts, then the large amounts of fluid deposited in the masonry through irrigation can mobilise existing inherent salts with the same damaging effects.

So we are always looking for more benign approaches to the use of treatments. By taking a holistic view of the impact of remedial work, we can seek to ameliorate damage through management, design, specification and craftsmanship. We want treatments to be minimal and therefore we were interested in some of Dr Ridout's early thoughts on research, particularly with reference to targeted effective treatments. We were also looking at ways to get behind the symptom of a problem to the absolute cause, and in the case of decay in structural timbers in cathedral roofs, we would argue that the deathwatch beetle infestation or the dry rot or the oak rot outbreaks are but symptoms of an underlying problem, which is water ingress into the fabric.

And so because of our disenchantment with disruptive, ineffective timber treatments, and through our ethical predilection for 'greener' approaches to the care of historic buildings, English Heritage moved towards the employment of risk assessment strategies for its conservation work. Here, we check for cause and effect; we undertake damage mapping and real time decay monitoring in order to evaluate the degree of risk to historic fabric before prioritising the way we then intervene. By such means, we are able to keep most of the old material for longer, by sensitive honest repairs. We make sure that our contribution to the welfare of the building is recognised, recorded and archived to inform future work.

CONTRASTING ATTITUDES TO THE TREATMENT OF FUNGI AND BEETLES IN BUILDINGS

In order to provide a broad context for this conference's technical discussions, I want to try to compare and contrast some of the difficulties English Heritage and others experienced before we started this research, with trends in the treatment of the *Serpula lacrymans* or the dry rot fungus. Through the pioneering work of Hutton + Rostron, and their former staff mycologists Drs Jagit Singh and Brian Ridout we have got to a situation now where we collectively understand the inter-relationship between moisture and the fungus. By using environmental control to monitor, control and limit the supply of water to the fungus, we can reduce its damaging impact without the need for excessive timber surgery or the profuse employment of fungicides.

As promulgated, these logical scientific trends towards non-destructive diagnostics, preventative conservation management and key-hole surgery related to fungal timber decay and its treatment do not seem to have influenced industrial thinking so far with respect to the deathwatch beetle. Only now, as the scientific findings from our Woodcare research are being delivered to practitioners and made relevant by technical advice on their application, should we now see a similar step change in methods and their effectiveness.

TIMBER REPAIRS TECHNIQUES

In the span of remedial treatments, the focus of today's discussions will inevitably centre on the monitoring and control of moisture; on the behaviour of beetles and the employment of traps and natural predation; on non-destructive diagnostics; and on the targeted use of pesticides within an integrated pest management strategy. However, there will still be a need for major structural interventions to make good damage caused by infestations that have led to missing construction or a lack of physical integrity.

So it would be remiss of the conference organisers if we had not bolted onto the end of the meeting some information and guidance on timber repair techniques to produce a rounded approach to the subject. There are different approaches to this subject: from the carpenters' craft tradition, using timber scarf repairs of various kinds;

to replacing missing elements with structural steelwork; and to the reinforcement of friable timbers with glass fibre or stainless steel rods or plates set in epoxide or polyester resin. Each of these techniques has its place in the conservators' lexicon. Each has its benefits and limitations. Unfortunately it has not been possible to publish these papers here. They are perhaps deserving of their own conference. However, advice on the subject is broadly available from existing publications (Faulkner 1965, Prudon 1975, Stumes 1979, Charles 1984, Ashurst & Ashurst, 1988) and should be read in conjunction with this volume.

If there are key messages to be learnt in this field they concern the general standard of building pathology in the United Kingdom and across Europe. Only by recognising phenomena in buildings for what they are, ie *symptoms*, and by searching for their true *cause*, can successful conservation take place. It is no good these days poking a screwdriver into a rotted beam end in the hope of gaining an insight into the scope and extent of a beetle infestation. That will not tell us very much. We need to know precisely where the beetles are, how many are emerging annually, whether they are juxtaposed with an outbreak of fungal attack, and all about the environment within which the beam and the beetles are placed, before deciding on a remedial response.

Our perceptions of abnormality in historic buildings are not technically aligned across the construction and remedial industries. Those surveyors, architect and engineers unused to dealing with large scantlings of ancient oak have a different perspective on their characteristics and capabilities than most of the specialists attending this conference. Our last speaker tomorrow will, I think, allude to this problem: some engineers are always looking for mathematical calculations to prove that structural members can function, whereas English Heritage staff regard historic buildings as the long-lasting field weathering experiments that, subject to monitoring, can patently be shown to work irrespective of the limitations of structural statics calculations.

The philosophical basis of our operations also needs to be aligned. If more people felt and operated in the way that English Heritage does, then there might be less damage to the historic built environment and more historic fabric left to pass on to the next generation. Valuable historic and archaeologically sensitive material in the marks of timber processing, the evidence of carpentry techniques, in decorative carving, jointing and decoration (Hewitt 1980) can all be lost if there is no skill in the assessment and evaluation of historic timber work. The marvellous work of previous generations can all be lost where poor diagnostic techniques limit responses to the crude, catch-all procedures of yesteryear. The effective use of diagnostics is as important as the treatment systems themselves (Fidler 1982).

THE EUROPEAN COMMISSION'S RESEARCH AND DEVELOPMENT PROGRAMME

Now I want to turn to the involvement of the European Commission and of our international partners in the collaborative research being shown today. English Heritage quickly realised that its resources alone could not hope to resolve all the outstanding scientific issues related to studies of deathwatch beetles in cathedrals. So in 1993, through the good auspices of Dr Ridout, it gathered together an international consortium to investigate the possibilities of undertaking collaborative research in the field. This cooperative venture included scientists from the United Kingdom, in the Department of Biology, Birkbeck College London, with the Jodrell Laboratory of the Royal Botanic Gardens, Kew; the Department of Wood Science at TNO-Bouw, the national building science establishment in Delft in the Netherlands; and the Departments of Chemistry and Biology at University College Dublin in the Irish Republic.

An application was made by English Heritage to the European Commission for a grant under its research and development programme in the fourth framework entitled 'Environment and Climate: technologies to protect and conserve the European cultural heritage'.[5] The consortia was fortunate to be awarded 500,000 ECU[6] for a three and a half year programme of activity, with the academic institutions receiving 100% costs and the two public bodies between 60% and 80%.

The research consortia involved not only mycologists, entomologists, chemists and wood scientists but also architects and surveyors representing the end-user specifiers and practitioners. It was particularly important from English Heritage's point of view that those who might employ the findings of the research should be involved in helping to steer and direct the scientific investigations from the start. Only by utilizing and understanding real life field situations would the scientists devise accurate testing protocols, hopefully to produce functional useful results.

English Heritage, of course, had several direct interests in the research. It was a building stockholder and had need of improved methods of dealing with beetle attacks on its own property. It wished to set new and improved standards of structural timber care for cathedrals and other historic buildings through its technical policies and grant aid. It also wanted to promulgate these new standards through its publications and outreach programmes.

Our research objectives were to study the interactions of deathwatch beetles and timber in historic roof spaces, noting that there seemed to be different interactions taking place between different ages of timber. We wanted to study the relationship of beetles to fungi (in particular *Donkioporia expansa*, the oak rot fungus) and between fungus and timber of different ages and between beetles, their natural predators and the environment.

As potential outputs from the research, we set down in our hypothesis at the start that we were to produce non-destructive diagnostic techniques to locate and target infestations and thereafter to validate the success of treatments. We wished to work towards trapping systems for beetles as a supplement to current chemical treatment, and finally to provide ecological data as feedback to the chemical treatment industry on the field predation model to describe what current chemicals are doing to our natural friends and allies in cathedral roof spaces, as well as to our enemies.

RISK ASSESSMENT AND INTEGRATED PEST MANAGEMENT

Ladies and gentlemen, in conclusion, as I have already inferred, our hypothesis has moved towards the concept of integrated pest management, dependent on local environmental conditions. Deploying a range of remedial responses to the beetles in their own environment in a targeted way can minimise disruption, save money and target treatments. But a precursor must always be a sound evaluation of the building envelope and its environmental conditions. Risks must thereby be enumerated and addressed accordingly. For example:

- If it can be established that an historic roof space is relatively dry, then the chances of having a major infestation of deathwatch beetle are limited and monitoring alone might suffice.
- In situations where there is some dampness, the chance of a beetle infestation increases. One can then start to adjust the moisture levels in the roof spaces by repairing the envelope and monitoring the drying process. By conserving, preserving and fostering the natural predator systems within the roof, or at least not disrupting them by targeted rather than broad-brush treatments, then beetle outbreaks can be controlled and a form of stability or equilibrium established.
- Of course in wet conditions, where because of the form of construction or lack of adequate maintenance there can be no guarantee of keeping the building envelope dry, there will be a need to use as many different systems as possible to help limit the impact of any infestation.

This subject will be illuminated in much greater detail and at length, I am sure, by my research colleagues here today.

CONCLUSION

Ladies and gentlemen, some of the scientific findings presented here for the first time today will be startling and immediately useful to practitioners. Others, equally useful, have been collated with additional historical and technical data to be published in a readily digestible form (Ridout 2000). Still others are definitely pioneering and promising but now need further research investment and development before commercial exploitation can ultimately benefit the field. The Woodcare research team and its associates do not have all the answers for specifiers and the remedial industry. Our findings, I am sure, move us all further along the road of understanding and progress for better building conservation for the century ahead.

ENDNOTES

1 English Heritage (the Historic Buildings and Monuments Commission for England) is a non-departmental public body, or quango, that is the lead body for the conservation of the historic built environment in the United Kingdom. It merged with the former Royal Commission on Historical Monuments for England (RCHME) in 1999, has around two thousand staff and runs a budget of over £140 million per year, the majority of which comes from central government by way of grant-in-aid from the Department for Culture, Media and Sport (DCMS). It has sister organisations in the rest of the UK: ie Cadw in Wales, Historic Scotland and the Environment & Heritage Service in Northern Ireland, carrying out similar functions.

2 Currently, English Heritage spends around £40 million per year on grants for the repair and conservation of listed buildings, scheduled monuments, conservation areas and historic parks and gardens. Between 1991 and 2001 English Heritage invested £34.5 million in its Cathedrals Grants Scheme. Included in this figure, for the financial year 2001/2, is the sum of £2.5 million for continuing work in this field. For example, York Minster will receive a grant of £113,000 towards the costs of masonry repair works in the Chapter House. Ripon Cathedral will receive £11,000 towards the roof repairs to its western towers.

3 The Grade I listed buildings targeted for special grant aid included the 43 cathedrals of the Church of England (most of them of medieval origin) plus 19 historic Roman Catholic cathedrals and significant non-conformist places of worship, making a total of 62 eligible sites. The total number of statutorily listed buildings of special architectural or historic interest in England in 1991 was then between 300–340,000 entries. Current estimates for 2000 place the total stock at over 430,000 entries in the statutory list.

4 The Cathedrals Research Client Liaison Committee (1991/2–1997/8) involved:

- Richard Halsey and Paula Griffiths, English Heritage's Regional and Assistant Regional Directors responsible for the cathedrals grants programme and its budget
- David Heath RIBA, English Heritage's Chief Architect (then Principal Architect for cathedral grants)
  Corinne Bennett ARIBA and Harry Fairhurst FRIBA, English Heritage's consultant architects for the cathedral grants programme
- Julian Limentani RIBA, Surveyor to the Fabric of Peterborough Cathedral, then Honorary Secretary of the Cathedral Architects Association and member of English Heritage's Cathedrals and Churches Advisory Committee
- Dr Richard Gem, Secretary of the Cathedrals Fabric Commission for England (CFCE)
  the staff of English Heritage's Building Conservation & Research Team.

Reporting to English Heritage's Cathedrals and Churches Advisory Committee (now called Places of Worship Advisory Committee) and the organisation's Science and Conservation Panel, the Liaison Committee helped to steer the relevance and programming of the applied research, monitor the budget and expenditure and peer reviewed scientific papers for publication, reporting upwards to English Heritage's Commissioners.

5 For details of all the European commission research programmes, see its website at http://www.cordis.lu/en/home.html

6 500,000 ECU (@ 1.3892 ECU = £1.00 at 1994 prices) = £360,000

BIBLIOGRAPHY

Ashurst J and Ashurst N, 1988 *Wood, Glass and Resins,* Practical Building Conservation Series, **5**. Aldershot, Gower Technical Press, 10–30.

Brereton C, 1991 *The Repair of Historic Buildings: Advice on Principles and Methods.* 2nd Edition, London, English Heritage.

Charles F W B and Charles M, 1984 *Conservation of Timber Buildings*. London, Hutchinson.

Department of the Environment and Department of National Heritage, 1994 *Planning Policy Guidance: Planning and the Historic Environment*. Planning Policy Guidance Note 15. London, HMSO.

Fairhurst H, 1991 *The Cathedrals Survey*, internal report, 3 vols, London, English Heritage.

Faulkner P, 1965 *Timber Work*. Ancient Monuments Directorate, Ministry of Public Building and Works. London, HMSO.

Fidler J A, 1980 Non-destructive surveying techniques for the analysis of historic buildings, in Marks S. (ed), *Transactions of the Association for Studies in the Conservation of Historic Buildings*. **5**. Kilmersdon, ASCHB, 3–10

Fidler J A, 1996 Introduction: strategic technical research in the cathedrals grants programme, in Teutonico J M (ed), *A Future for the Past: Proceedings of a Joint Conference of English Heritage and the Cathedral Architects Association 25th and 26th March 1994*. London, James & James (Science) Publishers, 1–5.

Fidler J A, 1998 Introduction to the Research Transactions series: historic building materials research at English Heritage, in Teutonico J M (ed), *Metals,* English Heritage Research Transactions series **1**. London, James & James (Science) Publishers, 1–5.

Hewitt C, 1980 *English Historic Carpentry*, London, Phillimore.

International Council on Monuments and Sites, 1966 *The International Charter for the Conservation and Restoration of Monuments and Sites (the Venice Charter1964)* Paris, ICOMOS

Phillips M and Selwyn J, 1978 *Epoxies for Wood Repairs in Historic Buildings*. Heritage Conservation and Recreation Service, Technical Preservation Services Division, US Department of the Interior. Washington DC, US Government Printing Office.

Prudon T H M, 1975 Wooden structural members: Some recent European preservation methods, *Bulletin of the Association for Preservation Technology*, **VII** (1), 26.

Ridout B, 2000 *Timber Decay in Buildings: The Conservation Approach to Treatment*. London. English Heritage and Historic Scotland with E & F N Spon Ltd.

Robson P, 1999 *Structural Repair of Traditional Buildings*, Donhead, Shaftesbury.

Stumes P, 1975 Testing the efficiency of wood epoxy reinforcement systems, *Bulletin of the Association for Preservation Technology*, **VII** (3), 2–35.

Stumes P, 1979 *The WER (Wood Epoxy Reinforcement) System Manual: Structural Rehabilitation of Deteriorated Timber*. Ottawa, Association for Preservation Technology.

ADDRESSES

Traditional structural carpentry repairs

McCurdy & Company Ltd: P McCurdy, Managing Director, Manor Farm, Stanford Dingley, Reading, Berkshire RG7 6LS, UK; Tel: +44 (0)1734 744866

Reinforced copolymer structural repairs

Beta International Trading BV: F Haasdijk, Managing Director, PO Box 128, NL-4920 AC Made, Eerste Industrieweg 1, NL-4921 XJ Made, The Netherlands; Tel: +31 162 672267; Fax: +31 162 672268: email: renocon@wxs.nl, www.amsterdam.nl/bmz/conserduc

Independent mycologists, entomologists and building pathologists

Hutton + Rostron Environmental Investigations Ltd, Netley House, Gomshall, Surrey GU5 9QA , UK; Tel: +44 (0)1483 203221; Fax: +44 (0)1483 202911

Dr Jagit Singh, Environmental Building Solutions Ltd, 30 Kirby Road, Dunstable, Bedfordshire, LU6 3JH, UK; Tel: +44 (0)1582 690187; Fax: +44 (0)1582 690188

Dr Brian Ridout, Ridout Associates,147a Worcester Road, Hagley, Stourbridge, West Midlands DY9 0NW, UK; Tel: +44 (0)1562 885135; Fax: +44 (0)1562 885312

AUTHOR BIOGRAPHY

John Fidler is a chartered architect and Head of Building Conservation and Research at English Heritage where he is responsible for technical policy development, scientific research, technical advice, standard setting, training and outreach programmes concerned with historic building materials decay and their treatment. He administered the European Commission's *Woodcare* research project; directed and chaired part of the international *Woodcare* conference and is responsible for English Heritage's production of technical literature including the Research Transactions series in architectural conservation.

Part I

Deathwatch Beetle

Deathwatch beetle and its treatment
Conclusions and some practical results from the Woodcare research programme

BRIAN V RIDOUT

Ridout Associates, 147A Worcester Road, Hagley, West Midlands, DY9 ONW, UK;
Tel: +44 (0)1562 885135; Fax: +44 (0)1562 885312; email: ridout-associates@lineone.net

This volume of these *Transactions* contains all that is known about the deathwatch beetle and the oak rot fungus at the end of the twentieth century. Undoubtedly the future, with advances in analytical techniques, will hold many surprises. Meanwhile it may be useful to summarise some of the practical implications from the Woodcare research programme.

When the Woodcare project began in 1995, we established a beetle hotline, which people could use to contact us and discuss their deathwatch beetle experiences. Thirty-five of the forty-eight people who telephoned us told the same story. They had deathwatch beetle, they had had deathwatch beetle treatment and several years later they still had deathwatch beetle. The beetles continued to emerge. 'Of course the treatment companies responded that the beetles pick up the chemical when they leave the wood, and then they die'. But imminent death did not seem to be on the agenda for the beetles that plodded across the carpet and so another imaginative argument was born. 'Yes, they are active, but don't worry, they have been sterilised and they can't re-infest'. But as the years turned into decades and the spring months still produced the familiar tapping behind the panelling, people became disenchanted with the excuses.

The standard method of beetle control was spray treatment and, according to the textbooks, this should have worked. Fisher (1938) had showed that eggs laid on the surface of timber hatched into larvae which wandered extensively before burrowing. Hickin (the Technical Director of Rentokil) enthusiastically deduced that the eggs could be laid some distance from the most favourable larvae feeding site, and their behaviour made the larvae particularly vulnerable to surface treatment (Hickin 1963). The eggs were vulnerable, the larvae were vulnerable and the beetles should pick up contact insecticides when they bit their way out of the timber. Unfortunately Fisher's observations and Hickin's deductions were based on an artificial situation. Fisher's beetles were presented with blocks of wood in glass jars, and their options were limited. The Woodcare project showed that beetle behaviour in buildings was rather different. Many beetles re-entered the timber after emergence, presumably to lay their eggs, and Fisher's wandering larvae were probably disorientated because they emerged onto a surface, rather than into a gallery in an appropriate environment. These conclusions need not surprise us because many creatures will eat an egg, and young larvae are particularly susceptible to desiccation. Why risk the vulnerable stages when the female beetle is robust enough to seek out a safe and suitable environment? The beetles only lay a few eggs and the welfare of all is important.

With the loss of two routes by which insect and insecticides might meet, we placed our faith in killing the emerging beetles, but this faith was misplaced. Many beetles do not bite their own emergence holes; they make their exit via whatever hole, shake or joint is available. Thus the results of surface treatments are always likely to be disappointing. Insecticidal smokes have produced better results, possibly because the surface area of contact with insecticide is much greater with dust particles, but nevertheless, and despite best endeavours, smokes have not eradicated deathwatch beetle infestations. We must conclude that surface treatments work best where there are no beetles.

But why should there be no beetles? This was the question that Maxwell Lefroy saw in 1924 as fundamental to an understanding of deathwatch beetles. If there are deathwatch beetles in a roof and the roof is made from timber, and the beetles eat timber, why should they abandon the task before the roof has been consumed? Fisher provided some of the answer when he showed that the insect larvae cannot attack oak heartwood unless fungus had also been present. Much of a roof therefore would not be available to the insects. The relationship between heartwood, fungus and beetle, however, eluded him. He and his colleagues thought that the fungus concentrated nitrogen, but this view cannot be substantiated, and we now believe that the fungus detoxifies the heartwood. The Woodcare research showed that a very small amount of fungus could radically alter the surrounding wood's chemistry.

These observations explain the distribution of beetles in a roof. The beetles may attack residual sapwood anywhere, but heartwood damage is likely to be restricted to core timber where there will be poor natural durability and possibly heartrots, or timbers which are decaying because of contact with wet walls. Nevertheless this restriction in distribution still does not explain Maxwell Lefroy's phenomenon, and we must seek an explanation in the environment.

There is plenty of evidence to show that the beetles flourish in wet (though not saturated) wood, but Woodcare also confirmed that the larvae can survive in quite dry conditions (moisture content > 10%). Nevertheless observations suggest that moisture is a limiting factor and large populations of beetles are rarely, if ever, found in dry timber. Disparate research on anobiid beetles suggested that dryer conditions prolong the growth period, producing smaller beetles which are less reproductively viable. Good maintenance for a prolonged period of time might therefore cause a beetle population to die back and fragment into small colonies which eke out their existence in the most favourable sites (timber bearings within walls, for example). This is doubtless an over-simplification but it does fit the generally observed pattern of beetle infestation.

Fragmented populations become increasingly vulnerable as increased isolation decreases population exchange, and factors such as natural predation become more significant.

An interesting study which perhaps parallels the death-watch beetle situation was made by Hill *et al* (1996) on the silver spotted skipper butterfly (Lepidoptera: Hesperidae) which colonises patches of short grass. They found that butterflies preferred to move between large patches of grass which were close together, and that small, isolated patches tended to become extinct.

The morals of this discussion are that good maintenance is essential, and natural predators should not be discouraged. Hickin stated that a major predator of the deathwatch beetle was a little blue clerid beetle *Korynetes caeruleus*, but Woodcare showed that house spiders might be even more significant. Clearly spiders are not going to have much of an effect upon a flourishing beetle colony, but they may seriously deplete small pockets of infestation and hasten them along the road to extinction. Incautious and untargeted treatments might not damage the deathwatch beetle population within the timber, but they will decimate the roving predators.

Preservative treatments must be targeted and the extent of the treatment must depend upon the magnitude of the problem. How is this all to be assessed? The easiest method is to find the beetles. Acoustic detection has now been developed but is not yet commercially available in the UK.

Deathwatch beetles emerge between April and June. In some buildings there is a slow build-up to a peak of emergence in May, followed by a slow decline, while in others all the beetles might emerge in a week or so. If deathwatch beetle infestation is suspected then treatment could be postponed until the following May, so that the number of emerging beetles can be assessed. A few more months won't make any difference. Any dead insects within the suspect area should be vacuumed up. This is of considerable importance, because a slow accumulation of dead beetles, perhaps over decades, can suggest a serious infestation when only a few beetles are actually emerging. It requires experience to judge the freshness of a dead beetle.

Woodcare found that the beetles will not fly unless the air temperature exceeds about 17°C. This means that in many buildings, particularly open churches, the beetles will tend to remain where they land when they fall from the timber, particularly if the floor temperature is below 10°C when the beetle will remain inactive. Many of these fallen beetles during the first part of the emergence season will be males that have fulfilled their purpose in life, and these might not fly irrespective of temperature. Female beetles tend to accumulate later in the season, and these may be more influenced by temperature.

The purpose of beetle collection is to locate pockets of infestation, and to access their importance. Beetles should therefore be searched for and collected throughout the emergence season and the numbers found in each location should be recorded. When the pockets of infestation are located then check to ensure there are no damaged gutters or other faults allowing water into the walls. If these are found then they should be remedied. Search for the source of the beetles. A fresh emergence hole will have sharp edges and an interior the colour of fresh-cut timber. There may be bore dust accumulating around it. Remember that many beetles will not bite their own emergence holes. Do not be surprised if there are many beetles but only a few fresh holes. If there are not any, but current infestation is still suspected, then fasten tissue paper or a similar thin paper tightly over groups of old holes, so that any emerging beetles have to perforate the paper. It is also worth noting that significant damage may have been caused by a long-standing infestation in structural timbers and an assessment by a structural engineer may be necessary. Concealed damage may be quantified with a decay probe, and hidden timbers may be located by thermography.

If emerging beetles are widespread, and hundreds of beetles are found, then total fumigation or heat treatment may be required. If beetles or frass accumulate below a few timbers then localized insecticide treatments will be appropriate, but these will have to penetrate well below the surface of the timber. Three options are available.

- Gels containing boron in a glycol base. These are useful if the moisture content of the timber exceeds about 20%. The formulations penetrate well, and the boron, if present in sufficient loading, will make the timber unavailable to the beetles. Boron is not a contact insecticide.
- Pastes formulated from mixtures of emulsions. These are useful if the moisture content of the timber is less than about 20%. They penetrate deeply into the timber and contain contact insecticides.
- Injection treatment. Insecticides are injected under pressure into the timber via embedded plastic one-way valves. Penetration of preservative along the grain is excellent because this is the natural route of moisture conduction. Penetration across the grain is more problematical, and many close-spaced injectors may be needed to obtain a uniform loading of insecticide. Many injectors may severely detract from the appearance of the timber.

All of these treatments have the disadvantage that the timber is a formidable obstacle between insecticide and insect. An alternative approach is to attract the beetles and deplete the population of its reproduction potential. UV light traps are successful, provided that they are in spaces where the air temperature exceeds 17°C, and are particularly valuable in churches with enclosed roofs. Care must be taken to ensure that the apparatus is bat-proof, because bats are attracted to the insects around the light.

BIBLIOGRAPHY

Fisher R C, 1938 Studies of the biology of the deathwatch beetle, *Xestobium rufovillosum* De Geer: Part II, the habits of the adult with special reference to the factors affecting oviposition, in *Annals of Applied Biology*, **25**, 155–180.

Hickin N, 1963 *The Insect Factor in Wood Decay*, The Rentokil Library, London.

Hill J, Thomas C and Lewis O, 1996 Effects of habitat patch size and isolation on dispersal by *Hesperia comma* butterflies: Implications for metapopulation studies, in *Journal of Animal Ecology*, **65**:6, 725–735.

Maxwell-Lefroy H, 1924 The treatment of the death watch beetle in timber roofs, in *Journal of the Royal Society of Arts*, **71**, 260–270.

AUTHOR BIOGRAPHY

Brian Ridout is a Director of Ridout Associates, specialising in complex damp and decay investigations, expert witness work, scientific research and lecturing. He has consulted for English Heritage for many years and was the technical coordinator and a subcontractor to English Heritage for the EC Woodcare project. He holds degrees from the Universities of Cambridge and London in entomology and mycology, and is a Fellow of the Institute of Wood Science.

Life cycle and feeding habits
Beetle behaviour in buildings and boxes

STEVEN R BELMAIN[1, 2], MONIQUE J SIMMONDS[3, *], WALLY BLANEY[2]

[1]Natural Resources Institute, Central Avenue, Chatham Maritime, Kent ME4 4TB, UK; Tel: +44 (0)1634 883761; Fax +44 (0)1634 883567; email: S.R.Belmain@gre.ac.uk; www.nri.org

[2]Department of Biology, Birkbeck College, University of London, Malet Street, London WC1E 7HX, UK; email: W.Blaney@bbk.ac.uk; www.bbk.ac.uk

[3]The Jodrell Laboratory, The Royal Botanic Gardens, Kew, Richmond, Surrey TW9 3AB, UK; Tel: +44 (0)181 332 5328; Fax: +44 (0)181 332 5340; email: M.Simmonds@rbgkew.org.uk; www.rbgkew.org

Abstract

Previous research on the life cycle of the deathwatch beetle, *Xestobium rufovillosum* De Geer, relied upon insects of unknown age and from non-standardized developmental conditions collected from infested buildings and trees. A culture of deathwatch beetle obtained from Germany enabled the comparison of the life cycle of the cultured insects in a standardized environment with those from buildings.

Adult deathwatch beetles emerged from the culture over a four to six week period. Males and females emerged in a 1:1 sex ratio. Adults lived approximately for one month. Females lived longer than males, unmated males lived longer than mated males and mated females lived longer than unmated females. Larval development (from egg to adult) in the culture wood varied from 1 to 13 years; the average time was five to six years. The duration of larval development could depend upon the variation and extent of fungal decay in the timber.

Deathwatch beetle females used cracks or insect emergence holes to place their eggs. This reduced the vulnerability of eggs to predation and chemical pesticides. Adults reduced their risk of predation by spending most of their time in cracks, crevices or old emergence holes. This also affected their susceptibility to surface pesticide treatments. Adults were mainly vulnerable to predation or to pesticides when they were more exposed through behavioural activities such as seeking mates or finding places to lay eggs.

Key words

Deathwatch beetle, *Xestobium rufovillosum* De Geer, timber pest, longevity, larval development

INTRODUCTION

Evidence of a long larval development time of ten years or more and a short adult emergence period of one to two months (Fisher 1937 & 1940) could be considered the primary disincentives for scientists to start a laboratory culture of *Xestobium rufovillosum* De Geer. Previous attempts at the Forest Product Research Laboratory (FPRL) to culture deathwatch beetles in the 1940s never met with success, and they were unable to maintain a viable deathwatch beetle culture (information from E C Harris).

The failures of these FPRL attempts were arguably due to extracting large larvae from willow trees and implanting these into pieces of oak timber via artificially drilled holes, according to the 'larval transfer test' (Hickin 1963a) as a means of speeding up culturing time. Transferring larvae resulted in high mortality levels. This mortality could have been caused through exposure to infection and dessication or to the inability of the larva to adapt to wood from different species.

Success in culturing the furniture beetle *Anobium punctatum*, with its shorter larval development time of two to three years, led to a standardized methodology and its widespread culturing for scientific experimentation (Becker 1942, Cymorek 1975). This methodology was successfully adapted by Michael Pallaske who started a deathwatch beetle culture in 1984 in order to evaluate pesticide activities. In 1995 the culture was transferred to the UK and is currently kept at the Royal Botanic Gardens, Kew. International enquiries suggest that this is the only deathwatch beetle laboratory culture in existence (information from B V Ridout).

The aims of the research presented in this paper were to compare the life cycle of the deathwatch beetle reared in culture with that of beetles collected from buildings, to discover potential effects of mating upon longevity, to establish adult emergence periods and sex ratios, eggs laid and to study larval development.

MATERIALS AND METHODS

Adult deathwatch beetles from the Pallaske culture were originally collected from infested buildings during the spring emergence period in 1984 (information from M Pallaske). Trunks or branches approximately 50–100 mm (2–4 in) in diameter were cut from oak trees, *Quercus* sp., in winter when the starch content was highest (Cymorek 1975). The wood was immediately debarked, cut into approximate 100 mm (4 in) lengths and then halved or quartered to give blocks of approximately 50 mm (2 in) in diameter. The blocks were quickly dried in an oven (300°C for 48 hours) to prevent the fixation of starch granules. The wood was autoclaved, allowed to cool and placed on a culture of the white rot fungus *Coriolus versicolor*, which had been grown on 2% malt agar

---

* Author for correspondence

Figure 1 Emerging Xestobium rufovillosum *adult. See also Colour Plate 1.*

extract for two weeks. The wood blocks with the fungus/agar then had an incubation period of eight to twelve weeks (21°C, 100%RH). After the fungal culture incubation period, the wood was again autoclaved and allowed to dry naturally for three days. Afterwards 0.4–0.6 mm diameter nylon fabric mesh, which helps to facilitate oviposition, was applied to one end of each wood block with methyl cellulose glue (Cymorek 1975). Four pieces of the prepared wood were placed into containers of 1–1.5 litre capacity, which were sealed with mesh to allow air exchange, and placed in a controlled environment at 21°C and 75%RH. Ten male and ten female deathwatch beetles were introduced into each container for every four pieces of wood. Containers were checked every second day, and any dead insects were removed. Beetles were collected daily on emergence and sexed according to Harris (1959) and Goulson et al (1993). As beetles were collected 50% were placed in containers to allow mating to occur, the rest were isolated in separate containers to prevent mating.

The culture was normally kept in a controlled climate environment at 21°C and 75%RH in total darkness. Artificial wintering, where the environmental conditions were changed to 7°C and 75%RH, commenced on 1st November every year and lasted 16 weeks, ending on 21st February. Wintering acted as a trigger for synchronization of pupation to encourage the adults to emerge synchronously in the spring (Cymorek, 1975).

X-rays were used to assess the density of larvae in the culture wood. X-rays have been used previously as a means of detecting wood-boring larvae in timber (Fisher & Tasker 1940, Bletchley & Baldwin 1962, Hickin 1963b, Baker & Bletchley 1966). In January 1997 three pieces of deathwatch beetle culture wood were X-rayed from culture years 1984 to 1996.

RESULTS

During the annual spring emergence period in 1996 and 1997, adult deathwatch beetles emerged from the culture daily. It took approximately one to four days for beetles to chew a hole large enough to exit the pupal chamber (Fig 1). The adult did not ingest the displaced wood and left a fine dust around the emergence hole. Females and males emerged in a variable daily ratio (approximately 1:1) with females slightly outnumbering males over the whole emergence period (Figs 2 and 3). The initial peak in numbers emerging in Figure 2 could be associated with the change in temperature at the end of the artificial winter one week before when the temperature was increased from 7°C to 21°C.

Mortality of emergent insects showed that mated males died much sooner than mated females (Mann-Whitney U-test = 44.18, n = 69, P < 0.005) (Figs 4 and 5) and unmated males were outlived by unmated females (Mann-Whitney U-test = 6.92, n = 69, P < 0.005).

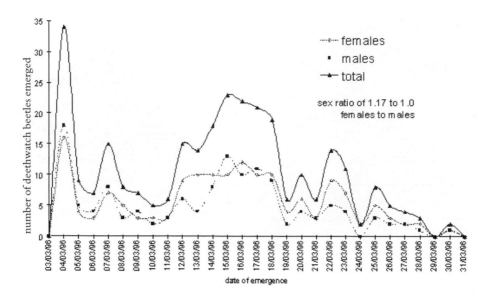

Figure 2 Daily emergence of male and female adult deathwatch beetles from the laboratory culture during the 1996 spring emergence period. Emergence profiles from 1996 and 1997 were similar (Spearman Coefficient = 0.999, n = 29, P < 0.0001); however, emergence of adults during 1997 lasted two weeks longer than in 1996 (Fig 3). In 1996 it took 11 days for 50% of adults to emerge, whereas in 1997 it took 16 days. During 1996, 294 beetles emerged, whereas 272 adults emerged in 1997.

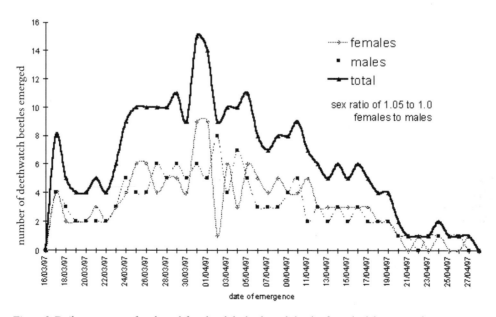

Figure 3 Daily emergence of male and female adult deathwatch beetles from the laboratory culture during the 1997 spring emergence period.

Unmated males and females had mortality profiles more similar to those of mated females than to those of mated males (Spearman Coefficient > 0.754, n = 43, P < 0.001) (Fig 4). Overall, mated males had the shortest life-span (±sem = 9.5±0.83 days) and mated females lived the longest (±sem = 32.7±0.19 days) (Fig 5).

Eggs oviposited by emergent females were observed in order to obtain information on the profiles of egg development time under standard conditions. The eggs took from 15 to 30 days to develop and the mean egg development time was 21.8±0.15 days (±sem, n = 416) (Fig 6). Eggs were normally laid in clusters of approximately 20 eggs and positioned in cracks and crevices.

The mean time-span needed for larval development under culture conditions was five to eight years from oviposition. During the 1996 and 1997 emergence periods, the first culture wood set up in 1984 still produced viable adults (Fig 7). During the 1996 emergence, wood set up one year earlier produced adult insects. Therefore, larvae in culture can take one to 13 years to develop in oak wood which had been exposed to *C. versicolor* under standardized culture conditions.

X-ray photographs of culture wood showed the death-watch beetle larvae as white objects in a range of positions dependent upon their orientation in the timber (Fig 8). The number of larvae in each wood sample varied.

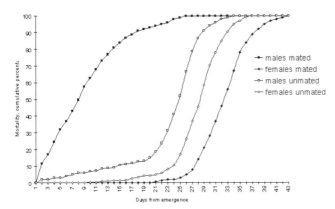

Figure 4 *Cumulative mortality (%) of deathwatch beetle adults from the laboratory culture (1996 emergence).*

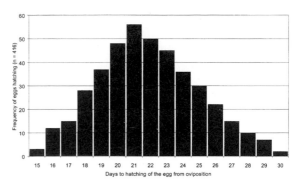

Figure 6 *Time taken for deathwatch beetle egg development, from oviposition to hatching (21°C, 75%RH).*

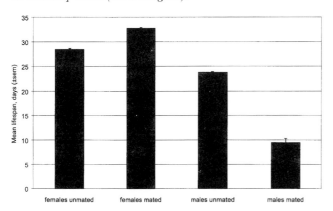

Figure 5 *Mean lifespan (days(sem) of deathwatch beetle adults from the laboratory culture (1996 emergence).*

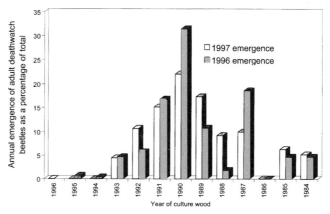

Figure 7 *Development time (years) of deathwatch beetle laboratory culture presenting time taken from oviposition to emergence of adults.*

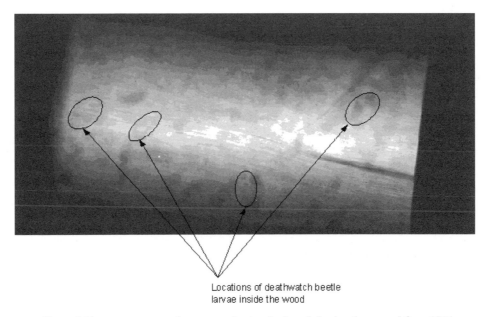

Locations of deathwatch beetle larvae inside the wood

Figure 8 *X-ray transparency plate generated using deathwatch beetle culture wood from 1991. Original transparencies show larvae more clearly.*

Despite this observed variance it could be seen that there was a trend for the density of larvae to decrease over time with the lowest number of larvae per 10 mm³ of wood occurring in the oldest cultured wood (Fig 9). With the exceptions of 1996, 1994 and 1986, larvae were clearly visible in all other years of cultured wood, including the oldest wood in culture from 1984.

Detailed observations on the laboratory culture showed that larvae, adults and pupae were found in pupal chambers in the wood during the winter months (November–February) in advance of the emergence season. Large larvae were observed to begin constructing pupal chambers in early summer. Detailed observations of larvae within the culture showed that it took several months for them to build their pupal chambers, which consist of frass pellets fixed in place by the larva. In most cases the deathwatch beetle pupated near the external surface of the host wood, within 1–5 mm.

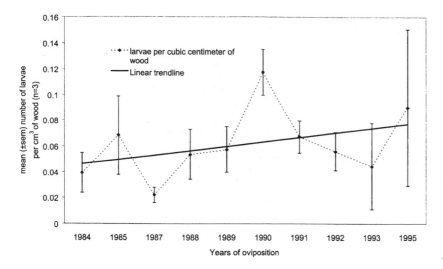

Figure 9 Density of deathwatch beetle larvae in laboratory-cultured wood determined by X-ray photographs of wood samples. Data from 1996, 1994, and 1986 were removed (see discussion for explanation).

DISCUSSION AND CONCLUSIONS

Larval development

This research showed a mean development time for the deathwatch beetle of five to eight years, from oviposition to adult emergence, which is shorter than the estimates of over ten years thought to occur in wild populations of beetles (Maxwell-Lefroy 1924, Kimmins 1933, Fisher 1937, Fisher 1940). However, these previous reports were not based upon controlled experimentation under standardized conditions.

Previous research showed that deathwatch beetle larval development was not associated with any particular fungal species attacking the wood, but larval survival was greatly enhanced by previous and/or concurrent fungal attack (Fisher 1941b). White rots and brown rots were found in timber infested with deathwatch beetle, and *Donkioporia expansa* in oak is often linked to deathwatch beetle infestations (Fisher 1940). The extent of fungal decay was positively correlated to the larval survival rate and negatively correlated to the duration of the life cycle (Campbell 1941, Fisher 1941a). With very decayed wood under high humidity between 20–25°C, the larvae emerged in 10–17 months (Fisher 1940). Maxwell-Lefroy (1924) reported that decayed heartwood and sapwood could support infestation. The exact nature of the interactions between the deathwatch beetle and timber fungi is unreported. For example, fungal decay could soften the timber, hydrolyze xenobiotics, improve nitrogen resources or make carbohydrates more readily assimilated (Jurzitza 1979, Buchwald 1986, Dowd 1989, Scrivener & Slaytor 1994). The roles of humidity and temperature in the development of the deathwatch beetle are also unclear. It has been shown that the larvae optimally required humidity above 60%RH, with 75%RH to 85%RH found most conducive to developmental requirements (Fisher 1940). However, deathwatch beetles can occur in buildings with very low relative humidities, going down to <30%RH with no apparent impact on beetle infestation, and the widely fluctuating temperatures that can occur in some building roofs would suggest

that the deathwatch beetle is tolerant of at least periodic high temperatures (>30°C).

Fisher's research on larval development was based upon an experimental methodology termed the 'larval transfer test' (Hickin 1963a) whereby extracted larvae were placed into artificially drilled holes in wood blocks. This methodology was initially devised to test pesticide efficacy. However, its use to determine larval development times was perhaps inappropriate. Discussions with the only surviving member of Ronald Fisher's original team, Ernie Harris, indicated that large larvae were extracted from willow trees and implanted into pieces of oak timber. This transfer usually resulted in high mortality levels, arguably due to infection and desiccation. As the adaptive range of *Xestobium rufovillosum* larval digestion is unreported, it could be that larvae from decayed willow would find it difficult to survive the transfer to relatively undecayed oak if, say, different enzymes are needed to digest different woods. Using insects that had already spent a large proportion of their lives in willow prior to the commencement of the experiment calls into question how Fisher determined the total larval development time relative to a wood sample. Despite many uncontrolled variables, Fisher's general conclusion that increased decay shortened larval development time could be considered likely, if indeed the insect's habitat was that of decaying trees. However, in the wild there must come a point when a tree is too decayed to support larval development through the reduction of resources by fungi and other deterioration. Elucidation of the precise details of decay effects on development would ultimately require insect rearing in culture from the egg stage on different types of decayed wood at different temperatures and humidities, and monitoring the emergence profiles over several years, perhaps even decades, to determine larval development time accurately.

Variability in larval development time shown in the culture where wood is all of a relatively standard decay level would suggest that such variance is caused by factors other than fungal decay levels affecting fitness, such as the number of larva per unit of wood mass. Research by Cymorek (1975) on *Anobium punctatum* showed that with a specific number of larvae, cooperative mechanisms

Table 1 Longevity of adults (days) under different conditions of temperature and humidity (Fisher 1938, 171).

%RH	15°C male	female	20°C male	female	25°C male	female	30°C male	female
23	36	46	19	24	17	19	14	19
41	44	52	21	25	14	21	14	16
53	42	56	22	28	20	22	16	21
75	52	56	24	34	19	29	15	23
95	36	63	27	49	22	35	15	34

could apply that diminish when the density of larvae increases, resulting in competition and cannibalism. Thus the highest relative production of adults does not necessarily take place with the highest number of larvae per unit of wood mass. Data on *Anobium punctatum* culturing suggested that a culture density of over three larvae per 10 mm³ of wood resulted in cannibalism and lower culture production (Cymorek 1975). As *Xestobium rufovillosum* larvae are more than twice the mass of *Anobium punctatum*, it could be argued that the density of *Xestobium rufovillosum* larvae should not exceed 1 larva/10mm³. X-ray photography of the culture wood blocks showed that the size of observed larvae within any given sample of wood varied considerably, suggesting that the rate of larval development was highly variable. The density of larvae within wood blocks was seen through X-rays to be lower than one would expect to cause competition for resources. However, the true causes of variance in larval development remain unknown. Although it has not been investigated, there is a possibility of density effects and superoviposition by *Xestobium rufovillosum* which could affect the number of hatching and developing larvae.

Emergence period

The emergence period of adults in culture was shorter (four to six weeks) than the reported emergence duration in buildings (three to four months) (Fisher 1937, Hickin 1963a). The more prolonged emergence in buildings may be due to the effects of more variable environmental conditions. Fisher (1937 & 1938) reported that adult emergence in buildings normally commenced in March. The duration of the emergence period in buildings varied, and was usually complete in the beginning of June (Fisher 1938). Living adults were rarely found beyond the month of July in buildings, even when the season had been relatively mild in temperature (Kimmins 1933, Fisher 1938). Later emergences in buildings have been reported (Linscott 1962). Other citations presented similar time periods for the emergence of adults in the spring (Maxwell-Lefroy 1924, Kimmins 1933, Hickin 1963a, Birch & Menendez 1991). Although no data were presented by Fisher (1938), emergence was thought to be temperature-dependent, requiring the ambient temperature to exceed 10–12°C in the spring for emergence to commence.

The minimum duration of a threshold temperature or the effects of the previous winter temperatures are unreported. Males apparently emerged in buildings several weeks before females (Fisher 1937, Harris 1964 & 1977). Previous research does not mention how emergence

periods were determined, and it is unknown if trapping or collection methods were used. It is unknown if previous research on adult emergence in buildings strictly monitored environmental parameters such as temperature and humidity, as no data were shown. Similarly, the number of buildings used to follow emergence profiles was unreported by Fisher or other researchers. The precision of determining the commencement and conclusion of emergence in a building is considerably more difficult in comparison to a laboratory culture as trap monitoring systems in buildings may not catch insects immediately after their emergence.

Life-span

The life-span of adult deathwatch beetles from the culture showed that 90% mortality in mated males occurred at 15 days from emergence, whereas 90% mortality in mated females occurred at 37 days. Previous research by Fisher (1938) showed life-spans ranging from 14 to 63 days according to changes in temperature and humidity (Table 1). He concluded that at any one humidity, an increase in temperature generally shortened the life of adults, but that while an increase in temperature from 15°C to 20°C shortened the life of the insects, a rise from 20°C to 30°C had little further effect on life-span. Adults were normally sluggish during the day and more active during the night (Kimmins 1933) and more active at higher temperatures (Fisher 1938).

Fisher did not discuss differences in life-span between males and females, although his data showed slightly shorter life-spans for males. As results from the laboratory culture showed that males and females emerged together, the difference in longevity between the sexes may explain the reports in the literature that male deathwatch beetles emerged first (Fisher 1937, Harris 1964 & 1977). The difference in mortality between the sexes would be seen when beetles were collected regularly in infested buildings as dead males would be collected before dead females, and this could give the impression that males emerged before females. In comparison to laboratory results, mortality effects in buildings or in the wild could be somewhat confused by lower, more variable temperatures and by variation in the length of time to mating and the frequency of mating, therefore blurring the distinctions between physiological states of insects trapped in buildings.

Longevity

Differences in longevity between the sexes could be argued to reflect the life cycle habits of deathwatch

Table 2 Duration of egg stage, days (Fisher 1938, 172).

Temp °C	23%RH	41%RH	53%RH	75%RH	86%RH	95%RH
15	No hatching (91)	49.7 (132)	41.3 (78)	45.6 (189)	–	44.9 (80)
20	No hatching (119)	23.6 (123)	21.3 (82)	20.01 (73)	20.7 (83)	21.1 (399)
25	No hatching (118)	15.3 (41)	15.7 (18)	14.2 (133)	15.7 (180)	12.5 (96)
30	No hatching (29)	No hatching (118)	No hatching (116)	No hatching (178)	No hatching (22)	No hatching (248)

*Figures in parentheses refer to numbers of eggs observed.

beetles. The adults do not feed but rely upon stored reserves (Fisher 1937), and it has been shown that the female deathwatch beetle demonstrated mate-selective behaviour based upon the size of the male (Goulson *et al* 1993). Research by Goulson *et al* indicated that male deathwatch beetles donated a large spermatophore, on average 13.5% of their body weight, to the female on copulation, the weight transferred being larger in heavier males and declining in subsequent matings. This large paternal investment could imply that the female uses the energy of the male's spermatophore as a food resource and to assist in egg production. The results on longevity from the culture are in accord with Goulson's work on paternal investment and female selection for large males as mated males would have less reserves after mating, leading to an earlier death, while mated females would have received energy, and would therefore be capable of living longer. Likewise, unmated males lived longer than mated males, the potential explanation being that unmated males have not had the opportunity to donate their spermatophore. Furthermore, unmated females do not live as long as mated females, the rationalization being that unmated females had not received the additional energy contribution from a male. It could be argued that deathwatch beetle tapping behaviour may, therefore, be fitness-related and dependent upon insect age, size or developmental health. Goulson's research showed that only virgin females would respond-tap, suggesting that as females appeared to mate only once, they selected large males to optimize their fitness and egg development. As little is known about the deathwatch beetle's natural habitat in forests, one can only speculate on why such mating behaviour evolved.

Oviposition

Fisher (1938) investigated various factors affecting oviposition, and his principal conclusions were:

- The commencement of oviposition did not appear to be affected by relative humidity. However, temperature did have an effect, whereby at 15°C the interval from emergence to first egg-laying was 18 to 21 days, at 20°C first egg-laying commenced at seven to ten days, at 25°C egg-laying started at six to 13 days, and at 30°C egg-laying started at six to nine days from emergence.
- The egg laying period was prolonged as the relative humidity rose and decreased as temperature increased. The duration of egg laying ranged from 17 to 35 days.
- The number of eggs laid per female increased at higher humidity from the normal 40–60 per female

up to 200+ per female. However, temperature had no apparent effect on the number of eggs laid.

Egg development

Fisher presented data on egg development across a range of temperatures and humidities (Table 2). He concluded that humidity above 41%RH had no effect on the rate of development of the egg, that the degree of dryness tolerated increased with increasing temperature, that temperature had a progressive effect on the rate of development with the minimum fatal temperature lying between 25°C and 30°C and the threshold egg development temperature lying between 10°C and 15°C. Viability of eggs under ideal conditions, ie low temperature and high humidity, was usually greater than 85% (Table 3). Although Fisher provided raw data on parameters affecting the egg stage, he did not show how these factors affected distribution patterns nor how these effects varied with insect age or mating status. Fisher's data on egg development were supported by research on development of eggs laid by insects from the culture. Eggs from cultured insects additionally showed that egg development time was normally distributed.

Pupation

The literature presents conflicting information as to when deathwatch beetle pupation occurs and how pupation is related to the date of adult emergence. Munro (1928) stated that pupation occurred in spring but that beetles did not come out until autumn or the following spring. Maxwell-Lefroy (1924) stated that pupation occurred from June to September, the insects emerging the following spring. According to Kimmins (1933), pupation occurs in late summer and early autumn, whereas Fisher (1938) believed that pupation starts in July and lasts about three to four weeks, the adults then over-wintering in their pupal cases to emerge the following spring. Birch & Menendez (1991) suggested that pupation could occur either in the spring or in the autumn. Whether these accounts were based upon detailed observations through the dissection of timber or speculation based on other research remains unknown. Although there are no de-

Table 3 Viability of eggs, % survival (Fisher 1938, 175).

%RH	15°C	20°C	25°C	30°C
23	0	0	0	0
41	85.2	61.3	73.2	0
53	91.3	95.9	87.4	0
75	97.0	92.8	76.8	0
95	86.1	96.4	90.5	0

tailed data on the subject, it may be that temperature dictates the synchronization of deathwatch beetle pupation, as has been shown for the furniture beetle (Cymorek 1975). If the trigger for pupation is tightly controlled by temperature it may follow that emergence times are actually dependent upon pupation time. However, if pupation is more loosely defined by temperature, it may be that insects pupate during the course of the summer and autumn as and when they have achieved the proper weight/instar (development stage). Research with the deathwatch beetle culture suggests that pupation and eclosion are somewhat loosely regulated by temperature because adults, pupae and larvae within pupal chambers were found within the culture wood during the winter months. In other words, some larvae over-winter in their pupal chambers, whereas other larvae pupate before winter, and some of these complete pupation before winter commences.

Detailed observations on insect pupation in culture suggested that the insect was capable of determining the external surface of the wood when boring brought them in contact with an edge of the wood block, arguably providing the insect reference for pupation near the surface. Reports of deathwatch beetles boring through lead roofing (anon 1931, Cathcart 1934) are likely to be related to adults emerging from their pupal chambers when timbers were positioned against the lead sheathing. Evidence in buildings showing that adults have emerged from old emergence holes (Ridout this volume), suggested that the emergence surface where insects pupated was not always an external surface but may have been an internal surface in severely decayed wood, for example, adjacent to a large cavity within a honeycombed piece of timber or a pre-existing insect tunnel. This could imply that deathwatch beetles may emerge into hollowed-out timbers, thus decreasing the risk from predation and pesticides as they complete their life cycle within a cavity in a timber.

SUMMARY

- Deathwatch beetles emerge in a 1:1 sex ratio. Their synchronization of emergence occurs with respect to temperature cues associated with the cold winter months.
- Adult longevity is sexually differentiated, males die before females. The act of mating further shortens the life-span of males and further increases the life-span of females. Longevity is also temperature-dependent; on average, adults live for approximately one month.
- The majority of deathwatch beetle eggs are laid five to ten days after mating. Females prefer to lay their eggs in cracks or crevices two to three microns across and can be placed at considerable depth within the timber, making the eggs less vulnerable to predation and surface chemical treatments.
- Larval development under culture conditions varied from 1–13 years to reach maturity; however, most insects in culture required between five to eight years to emerge as an adult.

BIBLIOGRAPHY

anon, 1931 Deathwatch beetles in church and other roofs, in *The Builder*, July, 180–184.

Baker J M and Bletchley J D, 1966 Prepared media for rearing the common furniture beetle *Anobium punctatum* (DeG.) (Col. Anobiidae), in *Journal of the Institute of Wood Science*, **17**:11, 53–57.

Birch M C and Menendez G, 1991 Knocking on wood for a mate, in *New Scientist*, 6 July, 42–44.

Bletchley J D and Baldwin W J, 1962 The use of X-rays in studies of wood-boring insects, in *Wood*, 485–488.

Buchwald G, 1986 Biological problems on *Donkioporia expansa* (Desm.), in *The International Research Group on Wood Preservation, 17th annual meeting, Stockholm*, (eds) Kotl and Pouzar, IRGWP.

Campbell W G, 1941 The relationship between nitrogen metabolism and the duration of the larval stage of the deathwatch beetle (*Xestobium rufovillosum* De G.) reared in wood decayed by fungi, in *The Biochemical Journal*, **35**, 1200–1208.

Cymorek S, 1975 Methoden und Erfahrungen bei der Zucht von *Anobium punctatum* (De Geer), in *Holz als Roh- und Werkstoff*, **33**, 239–246.

Dowd P F, 1989 In situ production of hydrolytic detoxifying enzymes by symbiotic yeasts in the cigarette beetle (Coleoptera: Anobiidae), in *Journal of Economic Entomology*, **82**:2, 396–40.

Fisher R C, 1937 Studies of the biology of the deathwatch beetle, *Xestobium rufovillosum* De Geer: Part I, a summary of the past work and a brief account of the developmental stages, in *Annals of Applied Biology*, **24**, 600–613.

Fisher R C, 1938 Studies of the biology of the deathwatch beetle, *Xestobium rufovillosum* De Geer: Part II, the habits of the adult with special reference to the factors affecting oviposition, in *Annals of Applied Biology*, **25**, 155–180.

Fisher R C, 1940 Studies of the biology of the deathwatch beetle, *Xestobium rufovillosum* De Geer: Part III, fungal decay in timber in relation to the occurrence and rate of development of the insect, in *Annals of Applied Biology*, **27**, 545–557.

Fisher R C, 1941a Studies of the biology of the deathwatch beetle, *Xestobium rufovillosum* De Geer: Part IV, the effect of type and extent of fungal decay in timber upon the rate of development of the insect, in *Annals of Applied Biology*, **28**, 244–260.

Fisher R C, 1941b Building reconstruction and the deathwatch beetle, in *The Journal of the Chartered Surveyors' Institution*, **20**, 3–8.

Fisher R C and Tasker H S, 1940 The detection of wood-boring insects by means of X-rays, in *Annals of Applied Biology*, **27**, 92–100.

Goulson D, Birch M C and Wyatt T D, 1993 Paternal investment in relation to size in the deathwatch beetle, *Xestobium rufovillosum* (Coleoptera, Anobiidae), and evidence for female selection for large mates, in *Journal of Insect Behaviour*, **6**:5, 539–547.

Harris E C, 1959 A note on the secondary sex characters of the deathwatch beetle. *Xestobium rufovillosum* De Geer (Coleoptera: Anobiidae), in *Entomologist Monthly Magazine*, **1140**:237, 208–209.

Harris E C, 1964 A field test of a lindane/dieldrin smoke for control of the deathwatch beetle, *Xestobium rufovillosum* (DeG.) (Coleoptera, Anobiidae), in *Bulletin of Entomological Research*, **55**:2, 383–394.

Harris E C, 1977 A long-term field trial of Gamma-HCH/ Dieldrin smoke against deathwatch beetle (*Xestobium rufovillosum*) in an ancient oak roof, in *International Biodeterioration Bulletin*, **13**:3, 61–65.

Hickin N E, 1963a *The Insect Factor in Wood Decay*, London, Hutchinson.

Hickin N E, 1963b *The Woodworm Problem*, London, Hutchinson.

Jurzitza G, 1979 The fungi symbiotic with anobiid beetles, in *Insect-Fungus Symbiosis: Nutrition, Mutualism and Commensalism*, (eds) Batra L and Montclair N J, Allanheld, Osmun & Co, 65–77.

Kimmins D E, 1933 Notes on the life-history of the deathwatch beetle, in *Proceedings of the South London Entomological Natural History Society*, 133–137.

Linscott D, 1962 A late emergence of *Xestobium rufovillosum* De Geer (Coleoptera: Anobiidae), in *The Entomologist*, **95**:1185, 58–59.

Maxwell-Lefroy H, 1924 The treatment of the deathwatch beetle in timber roofs, in *Journal of the Royal Society of Arts*, **72**, 260–270.

Munro J W, 1928 Beetles injurious to timber, in *Bulletin of the Forestry Commission*, 9.

Scrivener A M and Slaytor M, 1994 Cellulose digestion in *Panesthia cribrata* Saussure - does fungal cellulase play a role, in *Comparative Biochemistry and Physiology B-Biochemistry & Molecular Biology*, **107**, 309–315.

AUTHOR BIOGRAPHIES

Professor W M Blaney is an Emeritus Professor at the Department of Biology, Birkbeck College, University of London with expertise in insect physiology. He was the Birkbeck co-ordinator for the EU funded Deathwatch beetle project that supported the work presented in this paper.

Dr S R Belmain is a Senior Research Scientist at the Natural Resource Institute, Greenwich University, working on the control of insect pests of stored products. He was the Research Associate employed by Birkbeck College on the EU funded Deathwatch Beetle project. Part of the work presented in this paper contributed to his doctoral thesis.

Professor M S J Simmonds is Head of Biological Interactions at the Royal Botanic Gardens, Kew, a group that studies the role of plant- and fungal- derived compounds on insect behaviour. She assisted Professor Blaney in the coordination of the EU funded Deathwatch project and is currently working with Drs Brian and Elizabeth Ridout of Ridout Associates on the development of traps to monitor the beetles.

Finding a mate – tapping behaviour and deathwatch beetle communication, mate location and mate choice

Tristram Wyatt * and Martin Birch

Department of Zoology, Oxford University, South Parks Rd, Oxford, OX1 3PS, UK;
Tel: +44 (0)1865 270385; Fax: +44 (0)1865 270309; email: tristram.wyatt@zoo.ox.ac.uk

Abstract

Deathwatch beetles are named for their characteristic tapping behaviour. Both males and females tap, using their heads to hit the wood. Each tap consists of 4–11 head strikes at a frequency of about 11 Hz. The taps made by males and females appear to be the same. Males initiate tapping. Females only tap in response to the males. The beetles sense vibrations carried through the wood rather than sound: a short air gap is enough to break communication.

The males use the female's reply tap to locate her. Males move between taps whereas females stay still. The orientation mechanism the male uses appears to be a weak klinokinesis. Visual and pheromone cues do not seem to be important.

Females are more likely to mate with heavier males (probably because these offer larger spermatophores, which may be up to 22% of his body weight). Spermatophore nutrient resources are particularly valuable because the adults do not feed.

Key words

Deathwatch beetle, *Xestobium rufovillosum,* Anobiidae, tapping, vibration, paternal investment, sexual selection

INTRODUCTION

Surprisingly, the most famous feature of deathwatch beetles, *Xestobium rufovillosum,* the tapping which gives them their name, had never been described in detail until recently. In this paper we review the current knowledge of the tapping behaviour and how it is used by the beetle to find mates. We also describe some of the factors which may be involved in mate choice. A greater understanding of how male and female deathwatch beetles find each other may offer real opportunities for controlling this pest as part of an integrated pest management programme. Introductory accounts of deathwatch beetle behaviour can be found in Birch and Menendez (1991) and Birch (1998). More details of the experiments described here and biological background to communication in deathwatch beetle can be found in the references given in this paper.

In most wood-boring and bark beetles, long-distance communication is by pheromones (Borden 1997) and very close communication by sound or vibrations. However, unlike other species in the Anobiidae family, there is no evidence of chemical communication, using pheromones, in deathwatch beetles. Instead, tapping appears to be the only way male and female deathwatch beetles find each other. Pheromones have been searched for by behavioural and electroantennogram studies of deathwatch beetle male response to female ovipositor extract but no signs have been found to date (White & Birch, unpublished data). This is in sharp contrast with other important pest species in the family, such as the woodworm *Anobium punctatum,* in which males find females by responding to the sex pheromone stegobinone, released by 'calling' females (White & Birch 1987a, Birch & White 1988). The woodworm pheromone can be exploited for pest control, to detect woodworm by using traps baited with synthetic pheromone (White & Birch 1987b, Birch & Wyatt unpublished).

Although the larvae of the deathwatch beetle may not be detected for years the adults can be heard in the spring and early summer. The first descriptions of the tapping of the beetle date from seventeenth-century naturalists (Wilkins 1668, Allen 1698). Birch and Menendez (1991) speculated that the beetle got its morbid name in the days when the British usually died at home in their own beds. During the death vigil, or death watch, when the house was still, the tapping of the beetles was clearly audible. When a beetle taps, it sounds as though the noise comes from within the very structure of the house, and people regarded it as the countdown to death; the sound, perhaps, of the Grim Reaper, tapping impatiently with a bony finger waiting for the victim to die. This superstition was questioned as early as 1695, when Benjamin Allen reported in the Royal Society's *Philosophical Transactions* that 'yet I have known [the deathwatch] to be heard by many, where no mortality follow'd: and particularly by myself, who have taken Two of the same, Seven Years since, without any Death following that year' (Allen 1698). Allen's description of the beetle is sufficiently detailed for us to be sure that the beetle he studied was the same species that, in the British Isles at least, is known as the deathwatch beetle.

TAPPING BEHAVIOUR IS A RHYTHMIC COMMUNICATION

The tapping behaviour has often been considered as a 'sex call' (Fisher 1937, Fisher 1938, Hickin 1963) but the first experimental analysis to confirm this only came recently (Birch & Keenlyside 1991).

* Author for correspondence

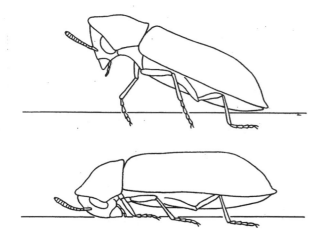

Figure 1 Diagram showing beetle striking substrate with its frons (Klausnitzer 1983, 115).

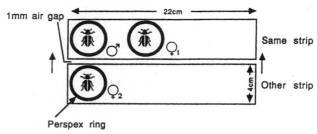

Figure 3 Experimental design to investigate whether signal is airborne or substrate-borne.

Both sexes of deathwatch beetle tap during their brief adult life. The beetles tap by repeatedly striking the wood with the frons or front of the head (Fig 1); each tapping bout typically consists of 4–11 strikes of the head, at a frequency of about 10 Hz (Fig 2) (Birch & Keenlyside 1991). The first task was to describe the tapping. Beetle taps were recorded using a modified gramophone cartridge with the needle touching the wood. The tapes were played back onto an oscilloscope screen or on a spectrum analyzer to allow analysis of the tap characteristics.

Airborne sound or substrate-borne vibrations

The deathwatch beetle gets its name from the sound of the taps which we hear. However, do the beetles communicate by these sounds, or do they detect vibrations passing through the wood when the beetle taps?

Small insects find it hard to make sounds that travel large distances. This is because the pitch (frequency) of sound that can be produced is related to the size of the sound-producing organ, so small insects can only produce high frequencies. In air these high frequencies attenuate rapidly with distance so do not travel far.

Producing sound in air is also a very energy-intensive activity. The solution used by many insects is to communicate with vibrations created by tapping the ground, leaf or tree. Their hard exoskeletons make ideal 'drumsticks' which send vibrations through the substrate when they hit their body against it. The substrate-borne vibrations can travel from leaf to leaf via the stem or along a tree trunk.

Do deathwatch beetles respond to the sound of the tap or the vibrations carried in the wood? Birch and Keenlyside (1991) tested this by placing a male and female deathwatch beetle a few centimetres apart on a pine strip (22 x 4 x 1 cm) (Fig 3). A second female was placed on an adjacent strip, positioned so that this female was slightly closer to the male than the first female. The strips were separated by a 1 mm air gap. Males were allowed to tap for one minute with the two strips separated by 1 mm and the females' responses were noted. The strips were then pushed together for one minute and then separated for a further minute, during which time the responses of the females to the male taps were again noted.

With separated strips, the female on the same piece of wood as the male responded to his tapping, but the second female, on the separate piece of wood, did not (Fig 4). When the strips were pushed together both females responded to male tapping bouts. Thus, it is very unlikely that airborne sound is important in deathwatch beetle communication, since the second female was at no time further from the male than the first female, regardless of the positions of the wooden strips. However, the role of airborne sound cannot be entirely excluded.

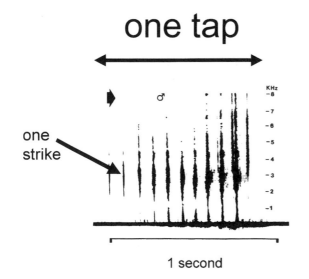

Figure 2 Sonogram of one tapping bout of one male deathwatch beetle (Birch & Keenlyside 1991, 258).

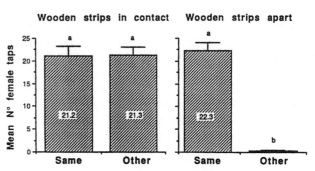

Figure 4 Tapping bouts of female deathwatch beetle (DBW) in response to male deathwatch beetle tapping bouts. Histograms show the mean (+/-SE) number of tapping bouts made by female deathwatch beetles, and those topped by a different letter differ significantly at P<0.001 (student's t-test). Wood strips together; n=30, wood strips apart; n=15 (Birch & Keenlyside 1991, 261).

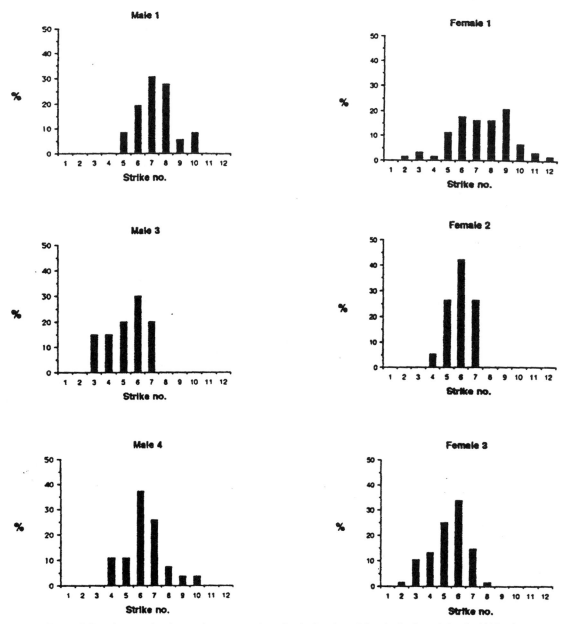

Figure 5 Distribution of strike numbers per tap for individual male and female deathwatch beetles. Note the variability between individuals (White et al 1993, 554).

WHO TAPS, WHEN?

Both sexes tap but characteristic male and female behaviour is very different. Males tap spontaneously and often, but females remain silent, only tapping in immediate response to another beetle's tap. Only unmated females respond. Active males and females establish tapping duets in which the male repeatedly initiates tapping bouts to which the female replies. Between taps the male moves, and the female usually remains stationary. The sequence ends when the male finds the female, probably by using her tapping replies as a means to locate her, and attempts to mate (see below).

CHARACTERISTICS OF THE TAPPING

Tapping therefore appears important for mate recognition and location in deathwatch beetles. Females need to be able to recognize male taps and respond to them, while males must recognize female taps and then locate their source. No differences were detected between male and female taps in the number of strikes, or strike frequency, per tap, although there was considerable variability in the number of strikes per tap (4–11) in both male and females (Fig 5) (Birch & Keenlyside 1991, White et al 1993). If there is apparently no difference between male and female taps, how do the beetles know if they have detected vibrations from a male or female tap? The clue may be the characteristic male and female tapping behaviours.

As females tend to remain silent, a male can ignore any spontaneous tap that does not closely follow one of his own taps. The time window seems to be about two seconds. A tap that he detects coming less than two seconds after his own is likely to be a female reply to his tap. It is just possible that it might be a tap by another male which by chance has ocurred then but if the first male establishes a duet, with a series of taps which each elicit a reply, then he must have established contact with a female

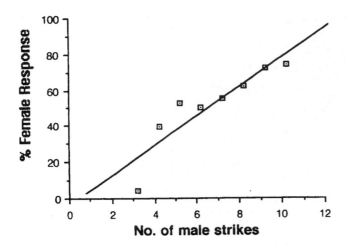

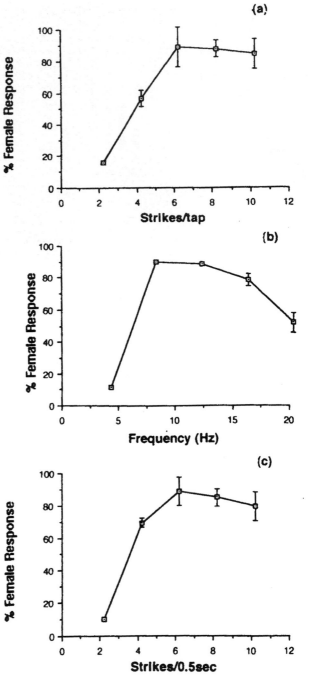

Figure 6. Summary graph showing female response to natural male taps with different strike numbers. Results are pooled for 330 taps from 15 male beetles to give the percentage female response to taps of each strike number (White et al 1993, 555).

(a second male would not respond to his taps). This then may be the 'rule of thumb' (tarsus?) that the beetles use.

Female taps may also enable males to assess receptivity, as females do not appear to tap after mating (Goulson *et al* 1993); thus a reply to a male tap tells him that there is a receptive female nearby.

EFFECTS OF MALE BEETLE TAP PARAMETERS ON FEMALE RESPONSE

Although the variation in the number of strikes per tap is similar in males and females, might this variability be important in communication? White *et al* (1993) looked at this in more detail by exploiting the natural variability in taps and using an artificial tapper.

An analysis of 330 male taps by 15 males, and female responses to these, showed that although there were no significant differences between the average taps of males and females, within each sex there were consistent differences in tap characteristics between individuals (Fig 5). In males there were significant differences between individuals in both mean strike number and strike frequency per tap. Mean strike number varied between males from 4.8 to 8.7 strikes/tap, while strike frequency varied from 10.0 to 12.0 Hz. Similar trends were found among females. Mean female strike number varied from 5.3 to 7.3 strikes/tap, while strike frequency varied from 11.2 to 12.2 Hz.

Females responded more readily to male beetle taps containing high strike numbers per tap (Fig 6). His strike frequency within a tap was not important, and his signal had no effect on the number of strikes or their frequency in her reply.

USING AN ARTIFICIAL TAPPER TO INVESTIGATE FEMALE RESPONSES

White *et al* (1993) were able to distinguish the features of male taps important to the female by using a mechanical tapper to make artificial taps similar to deathwatch beetle

Figure 7 Response of female deathwatch beetles to artificial taps varying in (a) strike number at a constant frequency (12 Hz), (b) frequency (Hz) at a constant strike number per tap (6 strikes/tap), and (c) both frequency and strike number at a constant tap duration (0.5 s). Each point represents the mean (+/- SE) of 10 individual females, each presented with 60 taps (White et al 1993, 557).

taps, but with a controlled strike frequency and strike number per tap. The tapper was constructed of a plastic disposable pipette tip glued at one end to a small loudspeaker (with signals coming from a signal generator), with the other end touching the wood. In the first set of trials, the strike frequency per tap was kept constant at 11 Hz, while the strike number was varied. In the second set of trials, both strike number and frequency per tap were systematically varied.

A higher percentage of females responded as the number of strikes in the artificial tap was increased, while

Figure 8 The tarsal pads of the deathwatch beetle may be the site of vibration detection (scanning electron micrograph; distance between tips of claw 0.1 mm).

keeping strike frequency constant at 11 Hz (Fig 7a). Females rarely responded to taps with less than six strikes. The artificial tapper allowed strike frequency, number and total tap duration to be varied. At a constant strike number, the percentage response of females rose with frequency, reaching a peak at 8 Hz, then declined at higher frequencies (Fig 7b). Again, the percentage response of females to artificial taps increased with strike number at a constant frequency (12 Hz), reaching a maximum at approximately 6 strikes/tap (Fig 7c). Taps of equal duration, achieved by altering both strike frequency and strike number simultaneously, showed an overall increase in percentage response to taps with a high strike number and strike frequency.

As females reply more readily to taps with six or more strikes, it is surprising to find males consistently producing taps with three or four strikes only (Fig 5). Such males should elicit few replies from females, so would be likely to locate and mate with few, if any, females. Sexual selection should act strongly against such males, so their presence in the population requires explanation.

Tapping characteristics may be related to male size, which varies considerably. Due to their higher inertia, larger beetles might tap at lower strike frequencies than smaller individuals, but there is no *a priori* reason to expect any effect on strike number. Beetle size is likely to have a significant effect on the amplitude of taps, but this aspect could not be measured with sufficient accuracy in these experiments and needs further study.

Male tap characteristics may also be age dependent. If tapping is energy intensive, older beetles could conserve energy resources by producing shorter taps with fewer strikes. In the tok-tok beetle *(P. striatus)*, which taps by striking the ground with its abdomen, however, tapping has been shown to be less energy-demanding than pedestrian locomotion (Lighton 1987). This suggests that energy conservation may not explain taps with few strikes in deathwatch beetle.

HOW IS THE SIGNAL DETECTED?

We currently do not know how the beetles detect the signal vibrations. One possibility is by mechanosensillae associated with the hairpads that form the pad of the tarsus of each of the beetle's six feet, between the hooks of the claw (Fig 8). Other possibilities are stretch or strain receptors, either in the cuticle or around the leg joints, that detect deformation. Clearly some interesting work remains to be done on this.

MATE LOCATION IN THE DEATHWATCH BEETLE

Characteristics of the female tap do affect the ability of males to locate an artificial tapper (and presumably a real female). A significantly higher proportion of males located the tapper at the higher strike numbers (6.0 strikes/tap) than at the lower strike number (1.4 strikes/tap) (White *et al* 1993). The tapping behaviour of deathwatch beetle clearly brings the sexes together by guiding the male to the female, but how does it work?

Mate location in deathwatch beetle represents a special case of animal orientation, since the beetles do not have a continuous gradient of stimuli to follow such as a concentration gradient (eg a humidity gradient), or a combination of pheromone and wind bearing from which the direction of the pheromone source can be inferred (Wyatt, in press). They cannot adjust their orientation by continuous sampling while moving, and the usual methods of track analysis are not applicable (Benhamou & Bovet 1992). Instead males must stop moving and elicit information at time intervals by tapping (Goulson *et al* 1994).

There are several possible mechanisms of male orientation to a female tap. Males may compare the timing or amplitude of vibrations reaching different parts of their body and turn towards the side with the stronger or earlier signal (tropotaxis). However, deathwatch beetles have a leg span of only approximately 2–3 mm which takes this to the limit of time resolution for the insect nervous system.

For these reasons comparison of female signal amplitudes from different positions is a more likely candidate as a basis for orientation. This may operate in at least two ways. If distance can be accurately judged from amplitude, then males could triangulate the position of the female, having received replies from three or more positions. Alternatively, turning rate may increase with decreasing amplitude of the female reply, so that if the male beetle is travelling away from the female it will receive a fainter signal and will tend to turn back towards it (klinokinesis) (Bell 1991).

More detailed experiments were carried out in the laboratory on flat horizontal arenas of solid hardwood, insulated from external vibration by support from expanded polystyrene blocks. In most experiments the artificial tapper was placed out of view of the beetle, underneath the wood, but tapping just 2 mm below the surface. Immediately above the tapper (and at three control sites on the arena), a dummy beetle (a recently frozen female) was placed as a visual target.

Male patterns of movement in the presence of artificial or real taps differed quite markedly from that observed in control experiments where there were no replies to taps. In the presence of 'female' replies (from a real female, or the experimenter via the artificial tapper), males tapped repeatedly from their starting position to establish a duet with a female (presumably to ensure that perceived replies were not taps coincidentally produced by another male at the time a reply was expected). Once a duet was established, the male began moving. The movement patterns were characterized by short straight walks at the end of which the male tapped, usually once, occasionally twice. The male then executed a turn before repeating the cycle. A sample track is shown in Figure 9b.

In the absence of female replies, males were either inactive, or travelled significantly longer distances between stopping to tap, tapped more from each position, and turned less at each position (Fig 9a). Analysis of male movement in the arenas suggested that males tended to turn more when further away from the female, perhaps as a response to the perceived decreased strength of the female reply.

To test for a taxis it is necessary to evaluate the ability of a beetle to orientate towards a female, having received replies from only one position (after receiving replies from more than one position other means of orientation are possible). Of 50 males, 28 made an initial turn away from the mechanical female, 22 towards. There was no significant bias of turns towards one end of the wooden strip (27 versus 23 turns). Males appeared unable to establish in which direction a female was positioned, having received replies from only one position, and thus presumably depend upon accumulating data by sampling female taps from more than one point to assess her location.

If males use klinokinesis to locate females, the angle of turning should increase with increasing distance from the female, presumably detected from the amplitude ('loudness') of her replies. No relationship was found between distance from the female and turn angle. Similarly, the target angle to the female did not decrease with successive movements: in fact there was a non-significant increase.

There was a significant increase in turn angle when the target angle was great, ie when the male was facing away from, and had just moved away from, the female. There was also an increase in the number of taps at any one position when the target angle was great. Lastly, the turn angle executed before leaving a tapping position tended to be greater when the male had produced more taps from that position.

When female replies ceased following a duet between male and female, males tapped repeatedly from the same position, presumably in an attempt to elicit a response. The males eventually executed a turn before moving. The turn angle was also greater than that found in males who received a consistent reply or no reply to all taps.

In summary, males do not appear to use orthotaxis during mate location, but exhibit a weak klinokinesis. Male turn angles tend to be larger following a movement away from the female, so that they turn back towards the female.

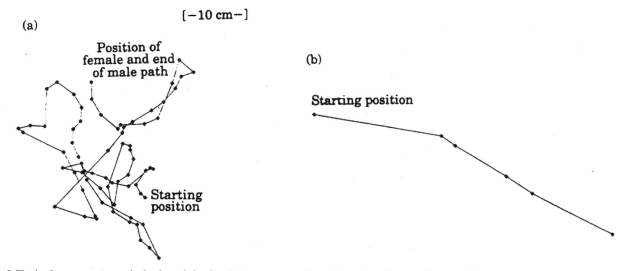

Figure 9 Track of representative male deathwatch beetles. Points represent places where the male stopped to tap, (a) with replies to taps, and female successfully located; (b) without replies to taps (Goulson et al 1994, 903–4, fig 3 for histogram of data analysis).

Male ability to locate females depends upon two mechanisms (Goulson *et al* 1994). The first is a simple switch between types of behaviour, according to whether a responsive female can be detected. If no female is perceived, males often remain inactive for long periods. When they do become active, walks are long between stops, and they tap many times at each stop. Presumably having established, by repeated tapping, that a female is not present in one area, there is little to be gained by stopping to tap again until the beetle has moved away from the barren area. Alternatively, if a female reveals she is nearby by her response to his taps, the male begins mate location behaviour. This is characterized by continual activity which consists of alternating short walks, tapping until a reply is received, and turns. The result is that the beetle tends to remain in the vicinity of the female. Clearly the increased activity of males in the presence of females serves in itself to increase the chances of finding a mate.

The second mate location mechanism consists of a means of orientating to the female tap. It is a type of klinokinesis in which turning rate increases following a movement away from the female, although not as a function of the absolute distance to the female. After walking away from the female, males tend to tap more and then to execute a larger turn. Male response to artificial cessation of mechanical replies strongly suggests that the movement away from the female is detected by a decrease in amplitude of the female response, as large turning angles can be induced regardless of movement relative to the female if replies cease. Presumably this situation simulates movement of the male beyond the range of the female reply, as would occur if the male moved too far from the female or crossed a small air gap in the substrate (the vibrations are not transmitted in air, Birch & Keenlyside 1991). When this occurs the male taps repeatedly in an attempt to elicit a response, and, when none is forthcoming, turns sharply: this serves to bring him back towards the female.

The mechanism described does not allow males to locate the female quickly, as demonstrated by the tortuous routes taken by males to locate females (Fig 9a), by the weak relationship between target angle and turn angle and by the number of males which failed to find the female before leaving the arena. Orthotaxis would be more efficient, but is perhaps not possible for an insect of this size (leg span approximately 2–3mm). Raleigh waves, the form of vibration detected by other arthropods (Brownell 1984), travel at approximately 3,800m s^{-1} in wood (Gogala 1985), and attenuate more slowly than other vibrational waves (Narins 1990). The time delay between waves arriving at different legs, allowing a generous legspan estimate of 3mm, is 0.8ms, which is too short for detection (Bailey 1991).

PATERNAL INVESTMENT AND FEMALE MATE CHOICE

Little is known about the mating behaviour of deathwatch beetle. Goulson *et al* (1993) provided the first

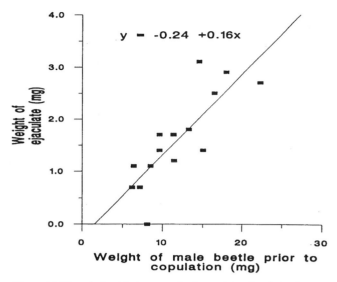

Figure 10 The relationship between the live weight of male beetles prior to copulation and the mass of ejaculate produced. Males transfer a substantial proportion of their body weight to the female during copulation (mean = 13.5%, range 0-22%, n = 15) in proportion to their weight (Goulson et al *1993, 544).*

experimental description of mate choice by female deathwatch beetle and in particular highlighted the very large investment made by the male when he transfers a significant proportion of his bodyweight in the spermatophore given to the female in mating (Fig 10).

Upon locating a female, males invariably attempted to mate by climbing upon the back of the female and extruding their paired valves. However, unless the female also extruded her genital apparatus from beneath her elytra, the male was unable to copulate, and generally abandoned the attempt within two to ten minutes. If the female did allow mating the pair remained in copula for 15 to 40 minutes.

Females reject many of the males that attempt to mate. The clear relationship between male weight and weight of their ejaculate suggests that females should choose heavier males. This is the case in laboratory experiments: successful males tend to be heavier. In other species spermatophores have been found to confer a number of nutritional benefits to the female. The donation of nutrients by the male parent is likely to be particularly important for egg development in deathwatch beetle, as their larval food, dead wood, is nutritionally impoverished. As females apparently mate only once, selection of a mate able to supply a large spermatophore is particularly important.

There is a suggestion only (Goulson *et al* 1993) that females might be more likely to accept a male if his weight had been artificially increased (eg with a blob of adhesive putty), but the numbers are too small to be sure. If females do judge males by weight, the likely moment is during courtship, when the male climbs on to the back of the female. If the male is not sufficiently heavy she does not extrude her genitalia and thus refuses to mate. However, these experiments need to be repeated before this can be established.

Although this study examines only female choice, given the large investment of males in each mating, male preference for large females may be predicted as these will lay more eggs. Also, female selectivity may vary with physiological state: for example, females may become less selective with age. Both provide interesting possibilities for further research.

DISCUSSION

Why deathwatch beetles rely on tapping for mate location while other related species, such as *Anobium punctatum* De Geer (White & Birch 1987a) and *Ptilinus pectinicornis* L. (Cymorek 1960), with similar wood-boring habits and ecology use female sex pheromones poses an interesting problem (see White *et al* 1993). Communication by vibration has been recorded from at least five widely separated beetle families (Crowson 1981). However, it is a comparatively rare means of communication, as more documented cases of communication for mate location in Coleoptera involve pheromones (Tumlinson 1985). Chemical communication has been described in several other members of the Anobiid family (Levinson & Levinson 1987, White & Birch 1987a, Birch & White 1988).

Tapping has far lower energetic costs than production of airborne sounds (Lighton 1987), and is likely to be less costly than pheromone production (Wyatt, in press). However, a key insight may be provided by *A. punctatum* which uses sex pheromones for distant mate location, yet taps during close range courtship (Birch & Wyatt, cited in White *et al* 1993). Common ancestors of modern anobiid species may have used both mechanisms. Males of *A. punctatum* tap as part of courtship when exposed to high concentrations of female pheromone. In this species vibrations may serve as an additional means of species recognition which operates at close range, and may avoid confusion with *Stegobium paniceum* L., which shares the same pheromone but does not tap (White & Birch 1987a). A similar case occurs in bark beetles, which stridulate when very close to each other as part of courtship, an additional barrier to interspecific mating. Again, communication over longer distances uses pheromones (Barr 1969, Rudinsky & Michael 1972).

Communication using substrate vibrations is less well studied than sound, perhaps because human senses are not well suited to detect them. Nevertheless, seismic communication, as this form of signalling has been termed (Narins 1990), has evolved in diverse organisms as varied as kangaroos, frogs, beetles and caterpillars. The feature common to all is that production of vibrations advertises the presence of the organism: it may also convey information on its location, species, sex and receptivity. The limited number of organisms adapted to detect vibrations is in itself a potential advantage of seismic communication as it reduces the number of predators or parasites which might otherwise be able to home in on signals intended for conspecifics (Narins 1990). Such parasite pressure could be another factor.

The behaviour of deathwatch beetle in male tapping, to elicit a female response and rapid moving on if none is received, is analogous to the 'call-fly' behaviour of male tick-tock cicadas, described by Gwynne (1987) and for a leafhopper, by Hunt & Nault (1991). Deathwatch beetles may use host wood cues to find good potential egglaying sites. Males arriving at these sites may use tapping to test if female beetles are there on the same tree or log, and then, if yes, use tapping duets to locate them.

Although various ideas exist for using the tapping behaviour for beetle control have been proposed, none is currently practical (Birch & Menendez 1991). One rather fanciful idea would be to create an artificial 'siren' female which would respond to tapping males and lure them to 'her', preventing them from finding real females. Another idea would be to set up continual vibrations through the timbers, at the same frequency as the taps, to mask the beetles' tapping and prevent mating. The trouble is that every susceptible timber, including church pews, would have to carry the vibration. The adult beetles are active for two or three months each year, and if the vibrations stopped for services, for instance, the beetles might seize the opportunity to mate.

CONCLUSIONS

Laboratory experiments have demonstrated the basic pattern of male and female communication by tapping, finally explaining the details of the sounds which give the deathwatch beetle its name. Other laboratory experiments have shown a fascinating story involving mate choice, which may have implications for control of this pest.

However, a key need revealed by these studies and others is for detailed investigation of deathwatch beetle behaviour in its natural environment, about which we know little, rather than in its adopted man-made habitat which has, in evolutionary terms, been available for a very short period. For example, the inefficiency of tapping as a means of mate location in laboratory experiments poses the question as to why tapping evolved instead of, or probably in place of, alternatives such as pheromones for communication. It may be that we are using the wrong experimental set-up for studying orientation behaviour. Only when we understand mate location and courtship in the beetles' natural habitat will we really understand its biology. This must be the next challenge.

BIBLIOGRAPHY

Allen B, 1698 An account of the *Scarabaeus Galeatus Pulsator*, or the Death-Watch, taken August, 1695, in *Philosophical Transactions of the Royal Society*, **20**, 349, 376–378.

Bailey W J, 1991 *Acoustic Behaviour of Insects*, London, Chapman and Hall.

Barr B A, 1969 Sound production in Scotylidae (Coleoptera) with emphasis on the genus *Ips,* in *Canadian Entomologist*, **101**, 636–637.

Bell W J, 1991 *Searching Behaviour*, London, Chapman and Hall.

Benhamou S and Bovet P, 1992 Distinguishing between elementary orientation mechanisms by means of path analysis, in *Animal Behaviour*, **43**:3, 371–378.

Birch M C, 1998 Death-watch beetles: undertakers of our churches, in *Antenna*, **22**, 195–200.

Birch M C and Keenlyside J J, 1991 Tapping behavior is a rhythmic communication in the deathwatch beetle, *Xestobium rufovillosum* (Coleoptera: Anobiidae), in *Journal of Insect Behavior*, **4**, 257–263.

Birch M C and Menendez G, 1991 Knocking on wood for a mate, in *New Scientist*, **1776**, 42–44.

Birch M C and White P R, 1988 Responses of flying male *Anobium punctatum* (Coleoptera: Anobiidae) to female sex pheromone in a wind-tunnel, in *Journal of Insect Behavior*, **1**, 111–115.

Borden J H, 1997 Disruption of semiochemical-mediated aggregation in bark beetles, in *Insect Pheromones Research: New Directions*, (eds) Cardé R T and Mink A K, New York, Chapman and Hall, 421–438.

Brownell P H, 1984 Prey detection by the sand scorpion, in *Scientific American*, **251**, 86–97.

Crowson R A, 1981 *The Biology of the Coleoptera*, Academic Press, London.

Cymorek S, 1960 Über das Paarungsverhalten und zur Biologie des Holzschäsings *Ptilinus pectinicornis* L. (Coleoptera: Anobiidae). XI International Kongress Für Entomologie, Wien, in *Verhandlange*, **2**, 335–339.

Fisher R C, 1937 Studies of the biology of the death-watch beetle, *Xestobium rufovillosum* De G. I. A summary of past work and a brief account of the developmental stages, in *Annals of Applied Biology*, **24**, 600–613.

Fisher R C, 1938 Studies of the biology of the death-watch beetle, *Xestobium rufovillosum* De G. II. The habits of the adult with special reference to the factors affecting oviposition, in *Annals of Applied Biology*, **25**, 155–180.

Goulson D, Birch M C and Wyatt T D, 1993 Paternal investment in relation to size in the deathwatch beetle, *Xestobium rufovillosum* (Coleoptera: Anobiidae), and evidence for female selection for large mates, in *Journal of Insect Behavior*, **6**, 539–547.

Goulson D, Birch M C and Wyatt T D, 1994 Mate location in the deathwatch beetle, *Xestobium rufovillosum* De Geer (Anobiidae): Orientation to substrate vibrations, in *Animal Behaviour*, **47**, 899–907.

Gogala M, 1985 Vibrational communication in insects (biophysical and behavioural aspects), in *Acoustic and Vibrational Communication in Insects*, (eds) Kalmring K and Elsnern N, Berlin, Paul Parey, 117–26.

Gwynne D T 1987 Sex-biased predation and the risky mate-locating behaviour of male tick-tock cicadas (Homoptera: Cicadidae), in *Animal Behaviour*, **35**, 571–576.

Hickin N E, 1963 *The Insect Factor in Wood Decay*, Hutchinson, London.

Howse P E, 1964 The significance of the sound produced by the termite *Zootermopsis angusticollis* (Hagen), in *Animal Behaviour*, **12**, 284–300.

Hunt R E and Nault L R, 1991 Roles of interplant movement, acoustic communication, and phototaxis in mate-location behaviour of the leafhopper *Graminella nigrifons*, in *Behavioral Ecology and Sociobiology*, **28**, 315–320.

Klausnitzer B, 1983 *Beetles*, Exeter Books, New York.

Levinson H Z and Levinson A R, 1987 Pheromone biology of the tobacco beetle *(Lasioderma serricorne* F., Anobiidae) with notes on the pheromone antagonism between 4S,6S,7S- and 4S,6S,7R- serricornin, in *Journal of Applied Entomology*, **103**, 217–240.

Lighton J R B, 1987 Cost of tokking: The energetics of substrate communication in the toktok beetle, *Psammodes striatus*, in *Journal of Comparative Physiology B*, **157**, 11–20.

Narins P M, 1990 Seismic communication in anuran amphibians, in *Bioscience* **40**, 268–274.

Rudinsky J A and Michael R R, 1972 Sound production in Scolytidae: Chemostimulus of sonic signal by the Douglas-Fir beetle, in *Science*, **175**, 1386–1390.

Tumlinson J H, 1985 Beetles: pheromonal chemists par excellence, in *Bioregulators for Pest Control*, (ed) Hedin P A, ACS Symposium, Washington, DC, **276**, 367–380.

White P R and Birch M C, 1987a Female sex pheromone of the common furniture beetle, *Anobium punctatum* (Coleoptera: Anobiidae): Extraction, identification and bioassays, in *Journal of Chemical Ecology*, **13**, 1695–1706.

White P R and Birch M C, 1987b 'An insect attractant for male common furniture beetles', UK Patent Application GB2 191 094 A.

White P R, Birch M C, Church S, Jay C, Rowe E and Keenlyside J J, 1993 Intraspecific variability in the tapping behavior of the deathwatch beetle, *Xestobium rufovillosum* (Coleoptera, Anobiidae), in *Journal of Insect Behavior*, **6**:5, 549–562.

Wilkins J, 1668 *An Essay Towards a Real Character and a Philosophical Language*, Royal Society, London.

Wyatt T D, (in press) *Animal Pheromones and Behaviour: Communication by Smell and Taste*, Cambridge University Press, Cambridge.

ACKNOWLEDGEMENTS

Thanks are due to the Association for the Study of Animal Behaviour and the Nuffield Foundation for funding, to English Heritage, and to the churches of North Aston and Great Milton where we collected the beetles. We also express our warm thanks to all those who have carried out many of the experiments and provided important insights, including Stuart Church, David Goulson, Chantelle Jay, Julian Keenlyside, Edwin Rowe and Peter White.

AUTHOR BIOGRAPHIES

Tristram Wyatt is University Lecturer at Oxford University's Department for Continuing Education and a researcher at the Department of Zoology. He took a first degree in Zoology and a PhD at Cambridge University before lecturing and research posts at the universities of Leeds, Berkeley California and Cardiff. He is a Fellow of Kellogg College, Oxford. His interests are in insect behaviour, particularly the ways that animals use pheromones or vibration for communication. He has worked on the pheromones and flight behaviour of predatory beetles used for biological control, woodworm beetles and storage pest insects.

Martin Birch has over the past thirty-three years worked in the fields of insect behaviour and chemical ecology. He was professor at the University of California, Davis before returning to Oxford as University Lecturer. He is now *ad hominem* research officer in Oxford University and Fellow in Lady Margaret Hall. He has used the manipulation of pheromones to study courtship in noctuid moths and as a pest control measure in pine and elm bark beetles, and artichoke moths. He has also worked on pheromones in woodworm beetles and vibration (tapping) communication in the deathwatch beetle in the anobiid beetle family. He has published three books on insect pheromones.

The population dynamics of the deathwatch beetle, and how their mode of attack influences surface treatments

Westminster Hall, London

BRIAN V RIDOUT

Ridout Associates, 147A Worcester Road, Hagley, West Midlands, DY9 0NW, UK;
Tel: +44 (0)1562 885135; Fax: +44 (0)1562 885312; email: ridout-associates@lineone.net

Abstract

This paper discusses the effects of moisture on the population dynamics of the deathwatch beetle by using Westminster Hall, London, as a case study. It also demonstrates how the mode of attack by the beetles is dependent on the presence of decay and how these modes of attack influence the efficiency of some surface-applied insecticide treatments.

Key words

Deathwatch beetle, heart rots, timber treatments, Westminster Hall

INTRODUCTION

In 1924 Maxwell Lefroy discussed a problem which he considered to be of considerable importance to an understanding of deathwatch beetle and its control. The problem was that beetle damage in buildings did not necessarily imply current infestation, and beetle populations could decline to a very low population density or become extinct even though there was ample apparently suitable timber available as a food source. Munro (1931) considered that the distribution of fungus within the timber might provide the key to the problem, and this is undoubtedly of considerable significance to beetle distribution and development (see Ridout & Ridout, this volume). Nevertheless it seems likely that the dynamics of a population will also depend on fluctuations within the environment. One of the most important factors will be timber moisture content, and some of the consequences of moisture content variation may be demonstrated by considering Westminster Hall as a case study.

Data from this building also give an insight into how a deathwatch beetle infestation proceeds, how the mode of attack is dependent on decay and how the mode of attack influences surface treatments with insecticides.

Westminster Hall, a building central in the history of England, provides what is probably a uniquely documented history of deathwatch beetle damage and its treatment. The Hall forms part of the Palace of Westminster and is currently maintained by the Parliamentary Works Office. It was here that the study of the beetle in the UK was initiated. Here too the first scientifically formulated insecticide mixtures were used to treat timber infestation in the UK, and over much of the twentieth century a wide array of chemicals were applied to the Hall's timbers to control beetle attack, yet there is little evidence to show that they were at all effective.

The beetle investigation commenced in earnest in about 1914, but Munro (1931) stated that only two live beetles could be obtained for study during 1914–18, in spite of the severe damage that was located. This damage provoked a massive treatment campaign which began with a particularly lethal insecticide formulation, but the beetle infestation was not eradicated. During the period 1951–61, for example, when daily counts were made of dead beetles which fell to the floor, the annual totals averaged 1900, and the total for the period 1951–71 was 43,849. The use of contact insecticides in spray and smoke form during the post-war years also failed to solve the problem, although the smokes undoubtedly killed beetles.

The beetle population (deduced from decreasing numbers of emergent individuals) did eventually decline again, but the causes of the cyclical pattern of attack and the contribution from insecticides require study as these problems are relevant to most other historical buildings where deathwatch beetle attack has occurred.

HISTORY

The building of Westminster Hall began in 1097 for William II, and the original oak shingled roof lasted for about 300 years. The current roof is of hammer beam construction, and was erected for Richard II during the years 1394–1400. It has been described as the finest open timber-framed roof in the world. The massive oak timbers were converted from trees felled in the King's Park at Odeham, the wood of the Abbot of St Albans at Bernan, and a wood near Kingston-upon-Thames. The roof, which is 72.5 m (238 feet) long, 20.6 m (67 feet 6 inches) wide (Fig 1) and rises to a height of over 27.4 m (90 feet) from the ground, was apparently constructed under the control of Hugh Herland, the King's master carpenter. Baines (1914) whose report provides the above information [1] calculated that it contained about 40,000 cubic feet of timber, exclusive of roof boarding.

Early repairs and alterations (1663–1819)

Baines (1914) stated that the earliest detailed records of repairs apparently dated from 1663 when the lead was

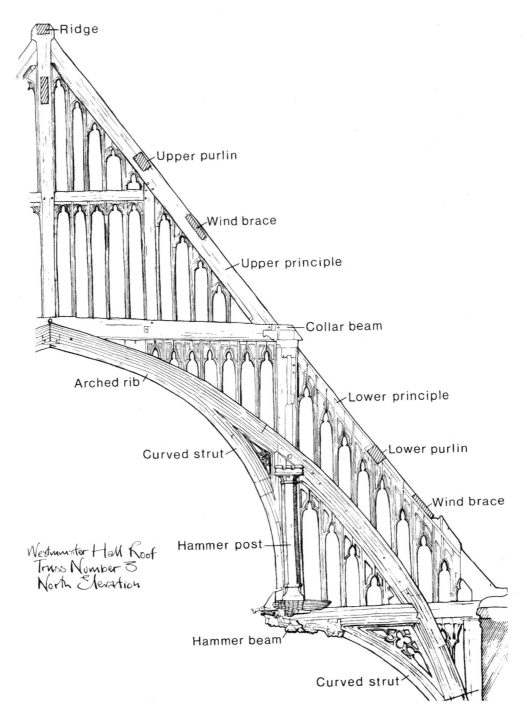

Figure 1 Section of part of the roof of Westminster Hall, showing components referred to in the text (after Baines 1914).

partially removed and recast, two hammer-beams and two collar beams were mended and pieced and three new purlins were inserted with iron bolts and stirrups. The gutters were also repaired and partially remade. In September 1663 the lantern was repaired and given extra support with additional timber. Five principal rafters were 'made good' with new timber and iron bolts, and three new plates plus one new rafter were installed. Three collar beams were similarly repaired. The leadwork was again removed at some time between 1760 and 1782 and for some unknown reason was replaced with slates. Other pertinent additions were the installation of hanging lamps in 1782 and two large stoves in 1793. The building appears to have been in poor condition at the beginning of the nineteenth century. [2]

1820–84

Blake (1925) quoted an article from the *Observer* (23 April 1821) concerning preparations for the coronation of George IV. When the roof timbers were drilled to suspend chandeliers it was found that the timbers concerned were so rotten that substantial replacement and repairs to joints had to be undertaken. Most of the oak within the rest of the roof was, however, reported to be in good condition, but the rafters, which were said to be cherrywood, were largely 'worm-eaten' and had to be replaced. Sir Robert Smirke commented in 1834 that the roof had been substantially repaired 'a few years ago' (Baines 1914) and we know that 12 dormer windows were installed on either side of the roof in 1819–20.

Munro (1931) stated that some of the roof timbers were repaired in about 1820 with sweet chestnut, but the presence of chestnut at Westminster Hall has never been confirmed. In 1850 Sir Charles Barry strengthened the roof with iron ties (Baines 1914).

1885–1908

On 24 January 1885 there was a dynamite explosion (attributed to the Irish Republican Brotherhood of Fenians) and the roof required re-slating. The dormers were also removed and substantial decay was found in adjacent timbers. [3] There are brief references in Baines to the following reports and repairs during the remainder of this period. None of the reports have been located.

- 1893 Report by an architect on purlins and corbels at the north end.
- 1897 Report by an architect on purlins and corbels at the south end.
- 1902 Report produced on gutters, parapet walls *etc.*
- 1906–7 Corbels at south end inspected and repaired (old repair remade 'in a proper manner').
- 1908 Four long rafters, lower purlin corbels and flashings were repaired at the centre of the west gutter.

1909–13

In 1909 two former sailors who had dusted down the roof timbers for many years reported a number of defects [4], and the lower purlin, plate and rafters up to the collar beam had to be renewed on the east side of the bay between trusses 4 and 5 (trusses are numbered from the northern end as in Figure 3). His Majesty's Office of Works therefore decided on a full inspection of the roof working on two out of the twelve bays each year.

The first two trusses to be tackled were those supporting the lantern. The timbers were said to be mostly in good condition with various previous repairs (1663) using timber and iron. The collar beams were, however, decayed on their 'inner faces'. This was attributed to lack of air movement, and the decayed areas were filled with steel plates and concrete. The latter method was controversial, and was discontinued. [5]

In March 1912 care of the building was transferred to the Ancient Monuments branch of the Office of Works, and it was decided, in view of the decay found and the lack of information as to the roof's condition and safety, to speed up the inspection process by scaffolding eight trusses. Substantial decay was located.

Samples of the damage, and of the organisms concerned were submitted to Dr Gahan at the Natural History Museum. The causes were described in the following manner (Baines, 1914, 10).

1. Dry rot
2. Incipient surface decay
3. Decay due to the attacks of the *Xestobium tessellatum* (one of the anobiid beetles)
4. Decay due to the attacks of a small anobiid beetle
5. The ravages of the goat moth.

Most of the damage was attributed to the deathwatch beetle then called *Xestobium tessellatum,* (now called *Xestobium rufovillosum*) but dry rot was seen as a major threat because of the new timber repairs, and lack of ventilation. It should be noted that at that period the term 'dry rot' covered any strand producing brown rot decay fungus, and it is unlikely that *Serpula lacrymans,* the 'true dry rot', was meant, because that fungus does not relish oak, although it is sometimes found attacking oak sapwood.

The smaller anobiid beetle is clearly the woodworm or common furniture beetle (*Anobium punctatum*) as no other anobiid beetle will attack the sapwood of oak. The 'ravages of the goat moth' (*Cossus cossus*) must in fact have been slight, because its burrows were only found in the sapwood of some common rafters. The goat moth is a pest of standing trees, and could not continue to infest building timbers. This is the only reference to goat moth attack in building timbers that has been located, and may be a misidentification.

The 'incipient surface decay' is of rather more interest, although not strictly relevant to this study. The architect W D Caroe believed it to be a surface stain, but Baines, a principal architect in the Office of Works, stated that it was 'cellular in structure' and that 'the fibre is perishing on (the timber) face' (Baines 1914, 10). Munro (1931), following the conclusions of Westergaard in 1914, said that it was a mould, and implied that it might be a food for young deathwatch beetles. This surface decay produced an ambivalent response when the restoration was discussed because it was also seen to impart a rich golden-brown tone to the timber, thought to be a distinctive feature of the roof. The cause was not resolved until 1983 when structural engineers from the Department of the Environment submitted samples of the surface degradation to the Building Research Establishment. Their verdict was that the damage had been caused by acid attack, probably sulphur dioxide from atmospheric pollution. This type of damage is actually quite common in London and other large industrial cities, and probably mostly occurred during the nineteenth century. In the present case fumes from the two large coke stoves installed in 1793 may have been a contributing factor. [6]

The results of the 1912 survey above were published in 1914 by Baines for the First Commissioner of Works. He reported that the roof was now in a dangerous condition and that considerable works were required, both to repair the damage and to destroy the beetles. His report is detailed, and provides the following summary of the types of damage found. Baines's sketch of truss 8, reproduced here as Figure 1, shows the location of each component.

BAINES REPORT FINDINGS (1912 SURVEY)

Hammer beams

Most were badly decayed, particularly where they joined the wall and the lower principal. Some had large cavities which had been repaired by inserting pieces of oak or by applying pieces of oak to the top surface. Iron shoes had also been used to provide additional support, and these, Baines concluded, seem to date the work to Barry's repairs of 1850. Baines goes on to say:

> The main decay manifests itself at the centre of the timber, and only when the centre is completely gone does surface decay appear to any very serious extent. It is somewhat curious that the decay occurs generally round the mortises for the tenons in the bottom of the post. (Baines 1914, 17)

Principal rafters

Most of the principal rafter feet were worm-eaten, some to a considerable extent.

> Some of the principal rafters would appear to have shrunk, and the mortises in the supporting members are exposed. In most of such cases the decay would appear to have started at this point, the cavities developing at the mortise holes, and running into both members where they join. This occurs also at the junction of the purlins with the principal rafters, where the mortises have been exposed by the sagging of the purlins and the decaying of the tenon. (op cit, 18)

Main collar beams

These are constructed from two timbers, each 305 mm (12 in) wide and 610 mm (24 in) deep, placed side by side. These two timbers had been bolted together at some period, probably the 1850s, but warping and structural movement had caused the timbers to part and deathwatch beetle had colonised from within the gap. Damage was, once again, serious at the joints between the collar beams and the associated timbers.

> In some cases the collar beam has been seriously attacked by the larvae along its whole length, and it would appear that the decay has commenced in such instances in the centre of the thickness of the two timbers extending into them and along them. (op cit, 19)

He stated that the beams most affected were those in trusses 4, 5, 8 and 9. The latter two were those filled with steel and concrete, as noted earlier in this section, and the decay was said to have continued behind it.

Hammer posts

These timbers are each 6.7 m (22 feet) long, and their width varies from 610 mm (24 inches) to 1.143 m (45 inches). They appeared to be in better condition than other timbers, but:

> The chief cause of the defects appears to be large dead knots where the larvae have worked, in one case, on the east side of truss No. 9 more than half the effective area of the post at its junction with the rib is thus destroyed. Dry rot would appear to be a contributory cause at this point, but the larvae have chiefly worked at the extremities of the hammer posts, particularly at the bottom ends. As described in the details of the hammer beams, the decay has progressed from the mortises into the tenons, thence working up into the posts. (op cit, 20)

The worst damage was apparently the total hollowing of the east post of truss 5.

Struts

These were found to be in reasonable condition.

> … there being only slight signs of decay, apparently due to sapwood. These timbers are throughout in quite a good condition, the free movement of the air about them having apparently preserved them both from dry rot and attacks of the larvae. (op cit, 22)

Wall posts

> These members, as a general rule, are in a dangerous state. Occupying, as they do, an unventilated position against the walls, and, in certain cases - particularly at the north end of the hall - being actually embedded in the wall and packed around with soft rubble, they have suffered very largely from dry rot. This, together with the action of the larvae, has so seriously affected the timbers that in almost all cases they are practically useless. (ibid)

New struts

These, together with other packing pieces, had been installed at some earlier date to support the failed wall posts.

> These new timbers, and also the feet of the struts in the wall, have been badly attacked since their insertion, the seating and packing pieces being practically reduced to powder. (op cit, 23)

Purlins

The upper purlins had been severely weakened by shakes and deathwatch beetle, but no details are given. However:

> … the middle or main trussed purlins are certainly stronger than any of the others. In some cases, however, the three members of which they are composed have been entirely hollowed out by the attacks of the *Xestobium tessallatum*, and instead of actually carrying the common rafters these purlins have been strengthened by having bolts fixed through them on to the backs of the common rafters.
> The lower purlins suffer from defects similar to those enumerated in regard to the upper purlins. They are, however, in a better condition throughout than the upper purlins, due primarily to the fact that many of them have been renewed. (ibid)

Wall plates

As a general note it may be stated that owing to their confined and unventilated position these members are usually in a seriously defective condition. Even when renewed, as many sections have been, decay due to dry rot, as well as to the attacks of the *Xestobium tessellatum*, has seriously weakened the timbers. It is questionable whether some means of ventilating the timbers at the wall heads should not be attempted, as at this point the decay of the wood appears to be very rapid. (op cit, 22)

Common rafters

Baines noted that these appeared to have been cut from small trees, or out of the outer section of large trees because there was sapwood present which had decayed. There is evidence that these rafters had been installed in 1820.

Ridge

This timber was said to be 'more or less sound throughout'. (op cit, 24)

REPAIRS

The 1912 survey emphasised the risk of roof collapse unless works were implemented, but there was no general agreement as to what those works should be. Sound sections of the roof were of considerable historical significance, so that the only options available were to remove the severely decayed material and resupport the rest with steel, or to undertake carpentry repairs with new oak. Baines considered that if the roof was resupported with a steel frame then only 35–40% timber replacement would be necessary. If, however, the timbers were to be repaired so that they were structurally adequate, then 70–80% of original material would be lost. Heavily repaired components could not act structurally as originally intended, and in some cases whole roof elements would have to be replaced.

The decision to use steel was furiously contested by those who thought that the entire character of the roof would be lost. The most vociferous of these was W D Caroe, then architect to the Ecclesiastical Commission. He considered that the approach was almost criminal and his objections were thoroughly voiced in the *Times* during 1913–1915. [7] Caroe claimed that the timbers had been inspected during the roof repairs of 1885, and that the only damage was the destruction of sapwood by worms, which had probably occurred 400 years previously. If the damage was as bad as claimed then it had occurred since 1885. This Baines knew to be impossible.

Caroe's comments were heeded, however, and on 15 April a Mr King brought the question before a House of Commons Committee. He stated that the question of why the damage caused by 'the beetle and the moth' had accelerated over the last few years had not been answered. He proposed two possible reasons for the increase, and

these are of interest. The roof timbers were now dusted and this allowed the beetles direct access to the timbers. The dust had therefore been protective. In recent years there had been a great many more prolonged sittings in the House during the winter months. He concluded the following: [8]

Why should that affect the roof of Westminster Hall? It is for the reason that if you have these timbers in a perfectly dry state the beetles ravages are stopped or modified, but when we have the House sitting and heated with hot air during the winter, then you have aqueous vapour and an amount of condensation on the roof of Westminster Hall and the timbers thereof that largely facilitate those insects.

All objections were eventually dismissed and the decision was taken to resupport the roof with a discreet steel frame. It was therefore essential that an effective programme of beetle control was formulated because a considerable volume of potentially infested material would be left in place.

TREATMENT

The eradication of the deathwatch beetles was a problem with no obvious solution, and so it was decided in 1913 that a committee should be formed. [9] This committee was to comprise entomologists from the British Museum (Natural History), the Imperial College of Science and Technology and a group of noted chemists and mycologists. [10] It was agreed that an insecticide formulation should be devised which would satisfy seven conditions:

- to be capable of destroying deathwatch beetle
- to be applicable to wood *in situ*
- must not appreciably affect the wood's colour
- must not materially increase flammability or cause a fire risk during application
- must not weaken, distort or affect the structure
- must not endanger the workmen applying it
- must be comparatively cheap because a large quantity is required.

One of the first suggestions, that of Professor Westergaard, Professor of Mycology at the Heriot-Watt College in Edinburgh, was to use a mixture of carbon tetrachloride and naphthalene, which was found to be highly effective in trials. This efficacy was, however, countered with the story of a hairdresser from Harrods. Apparently carbon tetrachloride had been used for a hairdressing process and the hairdresser had been overcome by the fumes and collapsed. They laid her down on the floor but unfortunately the chemical was heavier than air and so she died. [11]

Sir Richard Paget suggested that the timbers should be sprayed and injected with formaldehyde and then encased in gesso impregnated with potassium cyanide to keep the formaldehyde vapour and beetles within the timber. [12] This was objected to on the grounds that the gesso would be unsightly, but he responded that it could be scraped off or would fall off 'in the course of time'. The idea of a roof which rained plaster dust laced with cyanide is an interesting one.

Some ideas, for example that of a Dr Goabby [13] who suggested infecting all the larvae with a fatal epidemic, were impractical, while others, for example the use of coal tar distillation formulations, would affect the appearance of the timber. Eventually two serious contenders remained and these were Dr Dobbie (Laboratory of the Government Chemist) who obtained a liquid with excellent penetrating powers by passing sulphur dioxide over camphor, and Professor Westergaard with his naphthalene formulation. No agreement could be reached and so Maxwell Lefroy (Imperial College of Science and Technology, London) was asked to devise a formulation. [14] He proposed an exotic mixture based on tetrachloroethane and trichloroethylene. The ingredients of Lefroy's first formulation are shown in Table 1. [15]

Lefroy's original proposal provides the following rationale for each ingredient.

- Tetrachloroethane to kill the insects (intended as a fumigant)
- Cedarwood oil to protect the wood in the future
- Soap to hold the oil in and make the timber non-flammable
- Paraffin wax to hold in the oil and to prevent chemical action
- Trichloroethylene as a solvent and diluent; it is a feeble insecticide.

In 1924 Lefroy elaborated a little:

> The paraffin wax is a changeless substance, preventing dusting out, holding in the soap, coating the eggs, forming a definite film that beetles dislike; the soap is a definite poison, though harmless; the cedar wood oil was meant as a temporary local deterrent against infection from another part of the building, since the whole building could not be treated at once. (Lefroy 1924, 264)

The local deterrent was considered necessary because the length of the larval growth period was unknown. Gahan (Baines 1914) thought that it might be 12 months, and the beetle would therefore be a vigorous coloniser. In fact it was six to twelve years.

1914–23

The restoration works commenced in July 1914. Use of Lefroy's insecticide formulation was approved on 21 August, 1915 and he was also asked to supervise its application.

Timbers were brushed down and vacuum-cleaned, and the fluid was sprayed until the surface would accept no more, paying careful attention with special spray nozzles to difficult corners, joints and tracery. The apparatus used consisted of a 10 gallon acid-resistant metal container, mounted on an iron frame with wheels. It was provided with a hand pump and pressure gauge capable of reading to 120 pounds/square inch. Most of the task

Table 1 Lefroy's first formulation.

Tetrachloroethane	50%
Cedar wood oil	4%
Solvent soap	2%
Paraffin wax	2%
Trichloroethylene	42%

was accomplished at $413.7 kN/m^2$ (60 psi). Unfortunately the mixture nearly proved as toxic to the workman as it was to the beetles, in spite of the issue of special gas masks, and its use was discontinued. A modification was suggested by Heppells Insectox Laboratories as a non-poisonous alternative, and this was submitted to Lefroy for approval. [16] A letter from Heppells to Baines dated 24 November, 1917 gives the formula (Table 2).

The latter mixture did not contain a wax to hold the volatile dichlorobenzene within the timber, because the manufacturers considered that the soap would fulfil this role. Lefroy himself (1924) cast doubts upon its efficacy, in spite of his earlier approval, and suggested that the cedarwood oil had also been removed. He also stated, however, that because the roof was resupported with steel it did not really matter whether the mixture worked or not!

The work progressed slowly because of the First World War which restricted the supply of materials and personnel. A progress report on the works does not seem to have survived, in spite of Baines' intention to make sketches at every stage. Nevertheless it is possible to reconstruct the order in which the trusses were repaired from answers to questions on progress which were asked in the House of Commons. The chronology given in Table 3 emerges from the answers to the questions as supplied by Baines.

If we assume that each truss and bay was treated when it was complete, which seems reasonable, then Lefroy's formulation was actually only used on trusses 4 and 5 together with bays 4–5 and 5–6. This would be in accordance with Munro's assertion (1931) that Lefroy's involvement was not as great as is supposed, and that most of the credit should go to the Office of Works.

The extent of the treatments undertaken remains conjectural. It seems probable that the initial spray treatments supervised by Lefroy would have been thorough, but it is unlikely that he had anything to do with the project after trusses 4 and 5 had been completed. The cost of the restoration works increased far beyond the initial estimates, and Baines' report to the House of Commons in 1921 which provided the information in Table 3 was an answer to a suggestion that the project was now too expensive, and should be abandoned. Baines listed the various costings and showed how costs had increased, but he did not mention chemical treatments.

The work of restoration which had been commenced in July 1914, was finally completed, and the Hall was

Table 2 Heppells' formula.

Ortho para dichlorobenzene	91%
White Castille soap	7%
Cedarwood oil	2%

Table 3 The chronology of roof repairs, 1914–1922.

date	commenced		50% or more completed		completed	
	trusses	bays	trusses	bays	trusses	bays
commenced July 1914 (21.8.1915 Maxwell Lefroy's mixture approved)	4	4–5	–	–	–	–
25.3.1916			5	?	4	4–5
11.10.1917 (4.12.1917 Heppel's mixture approved)	6, 7, 8	6–7, 7–8	9, 10	9–10, 10–11	4, 5	4–5, 5–6
14.11.1919		6, 7, 8	6–7, 7–8, 8–9	9, 10	9–10, 10–11	
26.1.1921 Finished 1922	11, 12, 13	11–12	1, 2, 3	1–2, 2–3, 3–4	6, 7, 8	6–7, 7–8, 8–9

eventually reopened by the King and Queen with great ceremony in 1923. A direct result of Caroe's objections to the restoration seems to have been an obsession with the retention of original timber. On the 30 July 1915 Baines wrote to the secretary of HM Office of Works with assurances that only the minimum of original material was being sacrificed.

> I can give the most definite assurances that the work has been treated with an almost reverent supervision, and not an inch of timber is sacrificed for which I am not personally responsible, and for which I am not prepared to give an adequate reason. [17]

Sir John Baird, the First Commissioner of Works, stated that the original estimate of 35–40% timber replacement had been reduced to 10%. It was therefore inevitable that a considerable volume of decayed material remained.

1924–50

In 1931 A W Heasman of the Office of Works wrote to Fisher at the Forest Products Research Laboratory seeking advice on re-treating the roof. Fisher's answer, dated 12 September 1931, was that Heppells' fluid was still the best that was available at that time. Pressed for further advice on the possibility of current infestation, he wrote on 16 October 1931 that if only one application had been given during the 1914–23 restoration then the treatment was unlikely to have worked. [18] Proposals were therefore made to erect two scaffolding towers so that two trusses could be examined, but nothing was done. Then in 1935 the roof was scaffolded up to upper collar level, so that it could be cleaned for the Silver Jubilee Celebration. The opportunity could not be missed, and in May 1935 F R Cann of the Forest Products Research Laboratory inspected the timbers. His method was to search for beetles and their larvae, fresh exit holes and bore dust. No beetles were found in the roof or on the floor, or in sweepings from both sources. The only indications of recent activity that were found were beetle galleries that had extended into filler pieces which he considered had been inserted during the 1920s restoration works or earlier. The suggestion of an earlier date would be supported by observations that the beetles do not like fresh oak (Ridout &

Ridout, this volume). Baines (1914) also mentioned damage in packing pieces.

The conclusions of Cann's report are of considerable interest.

> Considering the roof as a whole the indications of activity at the present time were very slight indeed, in spite of the large number of timbers which were found to contain beetle exit holes. In view, however, of the unsatisfactory results obtained in recent years in the roof timbers of many buildings which have received one or two surface treatments with an insecticide, it would not be surprising if the deathwatch beetle were still active in Westminster Hall. [19]

Cann's conclusions were that the beetles might be still there, but at a very low population density. The treatments he referred to in other buildings would probably have been with Heppells' fluid or some other formulation akin to Maxwell-Lefroy's mixture.

On the night of 10 May 1941 a bomb penetrated the roof between trusses 9 and 13 on the east side. The ensuing fire took eight hours to extinguish. [20] War-time London had many problems to contend with but Westminster Hall was still high on the list, so that we find an Urgent Question being tabled before the House of Commons on 25 May 1941.

> Captain Strickland: Could steps be taken to protect the roof? There is a great gaping hole which cannot be doing any good to the structure.

On 19 July 1941 there is a note that scaffolding had been erected, debris cleared and preliminary arrangements had been made for the fixing of corrugated iron. The work was unfortunately further delayed by a labour dispute.

The actual damage to the timbers was described as widespread though not particularly severe, nevertheless the following estimate [21] for replacement timber was produced:

- Roof boarding 17.55 m³ (620 ft³)
- Common rafters 10.19 m³ (360 ft³)
- Purlins, ridge timber, wind braces *et cetera* 5.66 m³ (200 ft³)
- Tracer, moulded puncheons *et cetera* 15.56 m³ (550 ft³)
- The Flèche 31.13 m³ (1100 ft³)

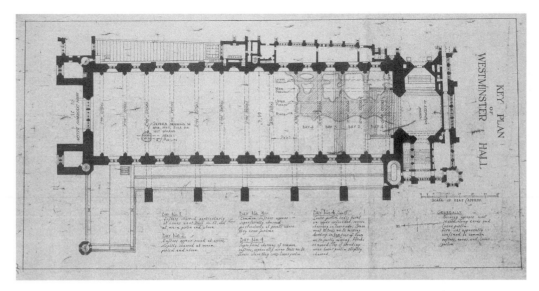

Figure 2 Plan showing wartime bomb damage (from Work 11.401.AE2509/40, Part 1).

Work was to be deferred until after the war. On 22 January 1944 two incendiary bombs fell through the roof. One disintegrated harmlessly on the floor, but the other lodged in the roof and caused a small fire which took 30 minutes to control. The damage covered an area of about 0.74 m² (8 ft²) and several timbers were charred, although not seriously. Both holes were covered with 'rubberoid felt'. [22] A contemporary drawing showing the extent of damage that was caused is shown in Figure 2.

Repair works eventually commenced in October 1947, but progress was slow because of a shortage of carpenters and because of trade union problems. These difficulties were finally resolved, and the roof repairs were completed during May 1950. A note states that all timbers accessible from the scaffolding were inspected for deathwatch beetle activity. 'Marked concentrations' of activity were dealt with by cutting out the damage, repairing and applying liberal insecticide. All new timbers were surface treated, probably with Cuprinol. [23]

1951–68

Daily beetle counts during the emergence season began in 1949 and were undertaken thoroughly from 1951 when repairs had been completed. The beetle count continued throughout the 17 year period. [24] Beetles were collected every morning, and their positions were recorded on charts. Enormous numbers of beetles were found emerging from the east end of truss 9, and in 1951 scaffolding was erected to hammer beam height. The upper part of the wall plate, and to a lesser extent the wall end of the hammer beam, were found to be heavily infested. The damaged sections were replaced, and the timbers were treated with 'liberal coats of insecticide', In 1952 the same treatment was used on the east side of truss 11. [25]

If the dots recorded on the daily charts are aggregated onto a year chart, then an interesting pattern of active infestation emerges (see Fig 3) The most consistently and seriously infested timber was truss 11 and the west end of truss 2. The pattern of beetle aggregation elsewhere seems to vary somewhat although trusses 10, 12 and 13 usually

produced beetles. The overall similarity of the charts suggest that the majority of beetles did not travel far from the timber from which they emerged. Recent research (Simmonds *et al*, this volume) has shown that the beetles are active flyers, so that some of the outlying spots may represent population dispersion, but temperatures during most emergence periods were probably too low for flight because of the thermal inertia of the huge hall.

The numbers of emerging beetles and the treatments employed since 1952 were charted in 1967. We have continued the chart to 1977 using the record sheets (Table 4).

1968–86

An interesting indication of the situation during the 1960s is given by a file note made by E C Harris of the Forest Products Research Laboratory. Trusses 8 and 9 were scaffolded for insecticide treatment. Truss 9 (west side) was carefully examined and it was found that the outer end of the hammer beam and the base of the hammer post had been badly affected. There was an outer sound shell 100–125 mm (4–5 in) thick and the interior of the beam was honeycombed with deathwatch beetle holes and old fungal decay of the 'wet rot' type. Damaged timber could be removed by hand leaving a cavity about 610 mm (2 feet) long which extended an unknown distance up the post.

> Only a few beetle exit holes were present on the surface of the wood near the affected area but one or two appeared to be of recent origin and living larvae were found within the cavity. [26]

Timbers were thoroughly treated, and the cavity was filled with new oak 'treated with an oil solvent preservative'.

The major interest in the Harris observations is that they closely resembles those of Baines made fifty years earlier. The fungus which caused a heart rot in the hammer post could not have caused the damage during those fifty years, so that we have an indication of the

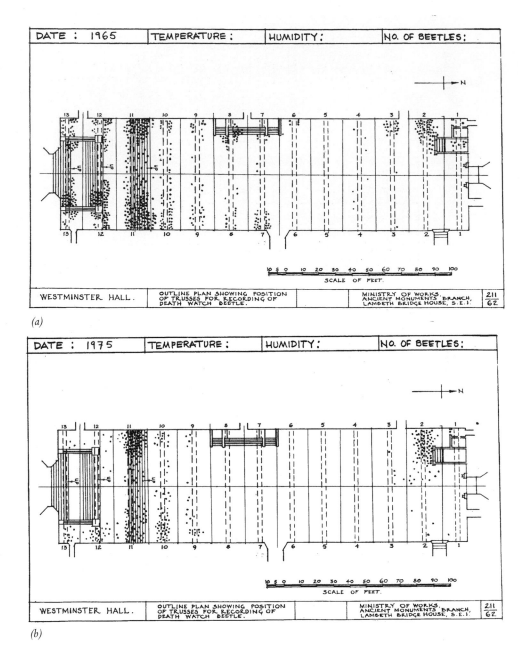

Figure 3 Distribution of beetles collected, (a) 1965 and (b) 1975.

extent of decayed (and hence original) timber which Baines was able to leave because of the structural steel framework which was inserted to support the roof.

By 1968 it had become apparent that surface spray treatments were having little effect, and smoke treatments were tested. These consisted of pyrotechnic formulations containing contact insecticides which were placed at intervals within the hall and ignited. The objective was to achieve a sufficient surface loading of insecticide on the timber to kill the emerging beetles. The 1968 smoke treatment was experimental and a rapid loss of smoke through roof vents was observed (Harris 1969). Surface deposits of gamma BHC (Lindane) were apparently measurable, but low, wherever tested. The exact quantity is not given in the report. These treatments, once established, were continued until 1979 when inter-

ference with the smoke alarm system caused them to cease. By that time, however, the annual total of beetles collected had dropped to about 400. In 1981 and 1983 the timbers were spray-treated with a water-based formulation containing the contact insecticide permethrin (Coopex WDP). No further treatments are recorded. The history of these treatments is given in Figure 4 and Table 5. [27]

Some of the sudden drop in beetle numbers from 1971 can be attributed to an imperfect collection procedure. In 1971, for example, the collecting did not begin until the annual emergence was well underway, and ceased while beetles were still emerging. In 1975 and 1977 the collector was on leave during the peak emergence periods. In later years beetle collecting was recommended at least twice every week, rather than every day, even

Table 4 The numbers of emerging beetles recorded between 1948 and 1977 and spray treatments applied to trusses.

WESTMINSTER HALL
Records of annual collections of deathwatch beetle
Compiled from data sheets sent by MOPB&W (Ancient monuments division)

Legend (cell hatching):
- Below hammer beam
- Cuprinol / Rentokil / Wykamol
- Entire truss

year	T1 E	T1 W	T2 E	T2 W	T3 E	T3 W	T4 E	T4 W	T5 E	T5 W	T6 E	T6 W	T7 E	T7 W	T8 E	T8 W	T9 E	T9 W	T10 E	T10 W	T11 E	T11 W	T12 E	T12 W	T13 E	T13 W	Annual E	Annual W	E+W
1948	97	8		3						5							576	4									676	21	697
1949	111	58															1366										1477	58	1535
1951	9	38	1	6							1	3	53	13	172	119	1322	22	209								1918	221	2135
1952	14	24	7	8	1	1		5	4	19	3	20	172	28	119	79	114	32	58	36	137	34	7	7		3	429	322	751
1953	9	43	3	22	2	3	1	6	6	44	20	20	28	12	114	61	40	46	63	29	31	30	2		1		358	306	664
1954	28	38	14	39	5	4	3	5	3	49	3	14	92	60	60	91	198	36	104	31	149	35	16		12		752	316	1128
1955	11	42	16	100	4	46	1	27	17	17	4	13	39	39	59	138	162	183	247	121	374	266	82	65	7	8	1126	909	2035
1956	23		4	1108	4	4	1	20	21	47	42	54	47		67	67	237	114	251	101	473	126		120		7	1207	718	1925
1957	5	139	5	123	5	7	17	17	17	16	2	1	32	1	32	3	173	51	57	21	455	41	57	19	17	7	856	329	1185
1958	1	8			1		1	1	1	1		1		9	43	43	266	135	344	126	433	59	33		75	17	1400	555	1955
1959	21	69	34	103	97	209	63	81	40	44	18		11	17	17	23	61	44	158	77	268	102	148	47	19	10	982	838	1820
1960		21	21	208		21	5	5	5	5	5	5	6	10	29	29	215	200	321	142	846	200	421	101	449	135	2131	1041	3173
1961		0	0	231	5	5	2	2		2	2	1	1	4	30	30	73	266	159	204	161	295	196	223	205	328	1256	943	2199
1962		173	91	91	3	3	2	3	1	7	7	9	12	15	30	30	249	153	325	253	887	605	201	34	141	9	1148	1359	2507
1963	1	2	1	191	1	3	1	3	1	5	5	5	45	31	55	36	221	219	228	226	330	306	256	266	276	205	1479	1490	2969
1964		9	16	239	2	1	2	7	16	10	34	34	33	33	48	36	170	172	163	147	230	182	173	153	247	163	1112	1169	2281
1965		9	5	109	5	34	5	5	0	5	2	10	51	15	12	74	27	47	141	84	376	249	258	69	141	110	1020	815	1835
1966	19	148	58	235	8	27					1		3	1	14	14	148	1650	148	171	171	174	256	205	247	128	1202	1290	2317*
1967	6	66	20	332	2	22	0	2	0	0	0	0	5	6	82	12	82	25	299	175	605	374	201	34	141	9	1364	1062	2426
1968		8	19	476	0	22	1	0	0	0	0	0	3	1	3	3	58	34	325	253	887	1150	154	12	161	3	1613	1964	3577
1969		38	0	431	2	13	1	0	1	0	1	1	1	13	1	13	38	38	258	228	754	799	163	40	208	24	1426	1638	3064
1970		23	31	281	0	22	0	0	1	0	0	0	12	12	8	8	18	32	186	170	742	739	122	42	174	11	1285	1329	2614
1971	1	11	6	122	6	6	0	0	0	0	1	0	7	2	7	4	34	9	140	44	198	459	71	32	80	28	529	717	1246
1975		2	5	140	2	8	0	0	0	0	0	0	18	2	18	1	18	7	126	22	103	193	38	8	19	14	312	392	707
1977	0	1	7	84	0	0	0	0	0	0	0	0	0	2	0	0	2	0	48	17	36	115	1	1	11	9	105	227	332

Note: some cells are shaded to indicate spray treatments (below hammer beam; Cuprinol/Rentokil/Wykamol; entire truss) as shown in the legend. Values in this dense, rotated table represent best readings and some figures are uncertain.

Table 5 *History of treatment 1968–1983.*

1968	Massive escape of smoke occurred (58 x Fumite No 40)
1969	Massive escape of smoke occurred (60 x Fumite No 40)
1970	No treatment
1971	Treated 4 April (100 x Fumite No 40)
1972	Treated 23 April, appreciable leakage of smoke occurred (80 x Fumite No 40)
1973	Two treatments 1 April and 6 May (each 80 x Fumite No 40)
1974	Two treatments 7 April and 5 May (each 80 x Fumite No 40)
1975	Not treated due to bomb damage previous year
1976	Two treatments 4 April and 2 May (each 80 x Fumite No 40)
1977	One treatment 15 April (104 x Fumite No 40)
1978	Two treatments 2 April and 7 May (104 x Fumite No 40)
1979	One treatment 8 April (104 x Fumite No 40)
1980	Not treated for security reasons
1981	Trial spray 0.25% permethrin wettable powder to the west ends of trusses 8 and 11 only (12 April)
1982	No treatment
1983	Treated with spray application of 0.1% permethrin wettable powder (5–10 April)
1984–1999	No treatments

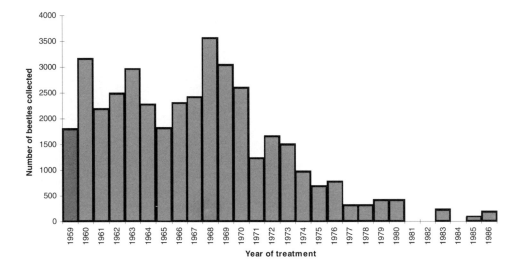

Figure 4 *Annual emergence of beetles from the Great Hall, Westminster.*

though the floor was swept every day. Nevertheless if we read daily totals for each year across Figure 4 then it seems probable that a drop in beetle numbers did occur during the early 1970s. The effects of the insecticide, and of environmental change, remain to be assessed.

DISCUSSION

Beetle populations, increase and decline

The lowest timber moisture contents which will support a deathwatch beetle infestation have not been ascertained. Some data from closely related beetles (*Anobium punctatum* and *Xyletinus peltatus*) are available, however, and these are shown in Table 6.

Information on a low population density deathwatch beetle infestation has been obtained by fastening tissue paper over clusters of beetle holes and measuring mois-

ture contents where activity is demonstrated (Ridout & Ridout, this volume). The results (Table 7) indicate that the beetles can tolerate quite dry timber. The building used in this study was in good condition, and had not been re-roofed for many years, thus suggesting a stable, but residual beetle infestation.

The effects of low moisture levels on wood-boring anobiid beetles have been investigated by several authors. Williams (1983) showed that a reduced wood moisture content retarded larval development in the American wood-boring anobiid *Xyletinus peltatus*. Meyer (1970) showed that a prolonged larval stage in *Anobium punctatum* produced smaller adults. Spiller (1948), again working with *A. punctatum*, demonstrated that smaller adults produced fewer eggs, while Goulson *et al* (1993) found that small male *Xestobium rufovillosum* were less reproductively successful with females.

Table 6 *Minimum environment required for larval growth in anobiid beetles.*

Species	Temperature	%RH	%MC	Reference
Anobium punctatum	20°C	65–70	12.0–13.5 (estimated)	Becker (1942)
Xyletinus peltatus	25°C	59.1±2.6	11.6±0.7	Williams (1983)

Table 7 Active beetle infestation and timber moisture contents at The Vyne 24.6.97.

Sensor number	Beetle holes 1996	Beetle holes 1997	%MC shallow 5 mm	%MC deep 20 mm
A1	0	0	14	15
A2	0	0	14	15
A3	0	0	13	14
A4	1	4	13	14
A5	3	0	13	15
A6	0	0	14	16
A7	0	4	10	12
A8	-	-	-	-
A9	0	4	14	15
A10	1	2	14	14
A11	1	0	12	14
A12	1	0	13	16
A13	3	1	16	12
B1	0	0	13	16
B2	0	0	14	15
B3	0	0	12	14
B4	0	0	13	15
B5	0	0	12	14
B6	0	0	11	14
B7	29	17	9	12
B8	0	0	12	14
B9	0	0	12	15
B10	0	0	13	14
C1	0	0	13	16
C2	0	0	11	15
C3	0	0	14	16
C4	0	0	14	16
C5	0	0	12	15
C6	0	0	13	16
C7	0	0	12	16
C8	0	0	14	15
C9	0	0	-	-
C10	1	2	13	15
D1	0	0	13	15
D2	1	0	12	16
D3	0	1	12	16
D4	1	1	13	15
D5	1	1	-	-
D6	1	1	13	16
D7	0	1	12	14
D8	0	0	12	15
D9	14	10	12	15
D10	-	-	-	-
D11	0	0	-	-
D12	0	0	-	-

Total number of sensors = 43
Number of sensors with beetle holes =16
%MC shallow = 12.6±1.59
%MC depth = 14.4±1.41
%MC shallow = 12.8±0.98
%MC depth = 15.0±0.77
Emergence holes from 1996 = 55
Emergence holes from 1997 = 50

If each observation applies generally to the anobiidae, then we may summarize these results by suggesting that low timber moisture contents extend the growth period of the larvae, and produce smaller adults which lay few eggs and are less acceptable to the opposite sex. Low timber moisture contents might therefore cause a beetle population to decline, and if conditions are consistently dry for long enough then the population may eventually become extinct. This might explain the observations made by Maxwell-Lefroy and others that there was frequently beetle damage without beetles even though

large volumes of apparently suitable timber were available. If this mechanism operates, then it is reasonable to assume that the reverse may be the case, and prolonged wetting might cause a population expansion. This is demonstrated by the Westminster Hall case study.

Mr King was nearly correct when he stated that dry timber could not be attacked by deathwatch beetle. King was, however, led astray by Caroe's assertion that the damage must have occurred between 1885 and 1911. The early repairs found during the 1914–22 restorations show that much of the damage probably occurred prior to 1850. In fact the few records available suggest an ongoing pattern of decay and repair. Munro and Maxwell Lefroy had difficulty in finding beetles; this is understandable because the roof was repaired several times during the nineteenth century, including re-slating in 1885. It seems likely therefore that the roof timbers were reasonably dry in 1914, and that most of the damage was historical, with a small resident beetle population surviving.

The dramatic increase in beetle emergence after the last war is easy to explain by elevated moisture contents. Bombs dropped in 1941 and 1944 caused fires, which were extinguished with water, and, more importantly, would have required widespread damping down (see Fig 2). The 1941 fire, in particular, was extensive and the timbers were exposed to the weather for two months. Eventually the hole was covered with a temporary corrugated iron roof. Research into buildings following fire damage (Ridout 2000) indicates three problems which may have increased the amount of moisture in the building.

- There would have been a severe condensation problem during the winter months because of the cold surfaces produced by the iron sheeting and scaffolding.
- The temporary roof could not have been made totally waterproof.
- The internal scaffolding would have wrecked the ventilation pattern of the hall.

This unsatisfactory environment persisted until May 1950, nine years later. We know from the 1935 inspection that the resident beetle population had probably survived the earlier treatment although there are no suggestions that the problem was severe, and the building was presumably dry. The population then had nearly a decade of ideal conditions following the fire damage in which to increase. If we compare the emergence charts with the fire damage chart and widen the zones of damage to include a generous margin for damping down by the fire brigade, then most of the subsequent pattern of emergence is encompassed, and we have a reasonable explanation of why the east ends of trusses 1–3 and the entire lengths of trusses 4–8 remained largely unaffected. These timbers were probably not damped down, and were therefore drier, though doubtless damper than they should have been because of the elevated ambient humidity.

Eventually, when the repair works were completed, the timbers started to dry down and the beetle population declined.

The modes of beetle attack

Baines' descriptions of the damage caused by the beetles allow us to categorize the modes of attack they exhibit.

Mode 1: Sapwood attack

Sapwood comes from the outer part of the trunk of a tree where water transportation occurs. Because it contains living material it has a higher nutrient content than other parts of the trunk and does not possess the natural preservatives (extractives) which some trees accumulate in their mature heartwood. The sapwood of any tree species is perishable when converted into building timber, and oak sapwood is readily attacked by deathwatch beetle larvae.

This form of decay seems to have been restricted by the minimal sapwood content of the timbers, and it is only recorded (though not specifically for deathwatch beetle) from the struts, of cross-section 171 mm (6.75 in) x 267 mm (10.5 in), and the common rafters, of cross-section 190 mm (7.5 in) x 152 mm (6 in).

Mode 2: Timbers hollowed by beetles with or without the presence of invisible fungal decay

- Hammer posts, of cross-section 914 mm (36 in) x 610 mm (24 in)
- Hammer beams, of cross-section 533 mm (21 in) x 571 mm (22 in)
- Principal rafters, of cross-section 419 mm (16.5 in) x 292 mm (11.5 in)
- Collar beams, formed from two cross-sections each 483 mm (19 in) x 301 mm (12 in)
- Trussed purlins, of average cross-section 356 mm (14 in) x 229 mm (9 in).

The centre of a tree contains the juvenile core wood. Juvenile wood comprises the first 10–20 years of growth and the deposition of natural preservatives is frequently small or non-existent. This is the most vulnerable part of a living tree because the sapwood is too wet to attack. If fungi can enter the juvenile core of a living tree via root or branch damage then decay can begin, and eventually the trunk will be hollowed. The risk that there was incipient and undetected heart rot in the centre of a building timber increases with its cross-section. Incipient decay in exceptionally large section timbers may continue to develop for many decades after the component has been installed because of the length of time it takes the component to dry. This will provide an ideal environment and food source for the beetle. The timbers listed above are all large section, and although fungus is not recorded necessarily from their centres, it was probably present (Ridout & Ridout, this volume). Baines reported the presence of 'dry rot', particularly in the centre of the hammer posts at their bases. This description indicates a brown rot (not 'true' dry rot), and the position of the hammer posts away from sources of damp suggests that the decay occurred before the timber was used in the building. It was therefore probably due to a heart rot in the standing tree. Likely candidates for the fungus would be *Fistulina hepatica* or *Laetiporus sulphureus*. The fact that the fungus was presumably not detected at the time of construction and continued to cause decay after installation of the beams, suggests *L. sulphureus*, because attack by *F. hepatica* usually ceases when the timber is converted (Cartwright & Findlay 1946).

Beetles probably seek out and attack the core of the timber because decay, juvenile wood with a low extractive content or a more stable environment is likely to exist there (Figures 5 and 6). They cannot attack the heartwood of oak unless decay is also present. Baines notes that the beetles appeared to enter the timber via gaps at joints, or between the two elements of the collar beams. Mark and recapture experiments during the Woodcare Project support these observations (Figures 9 and 10).

Mode 4: Timbers decayed by wet rots, because of direct water movement

Timbers that made direct contact with the walls had been decayed by fungus irrespective of their size, and these components included plates, wall posts and new strut ends. Baines attributed this decay to lack of ventilation, but it was certainly caused by prolonged contact with damp masonry. The type of fungus concerned cannot now be identified, although the description 'dry rot' suggests a brown rot. This would be a decayer of dead and wet wood, rather than a heart rot originating in a living tree. Baines' descriptions indicate that the combination of fungus and deathwatch beetle was particularly destructive. This is demonstrated by the replacement of the wall plate and the repair of the associated hammer beam end at the east end of truss 9 in 1951 when the number of beetles collected dropped from 1322 in 1951 to 114 in 1952.

The effects of surface treatments

Reinfestation from elsewhere in London seems highly unlikely, so that we may state with reasonable certainty that the Lefroy/Heppells treatment did not eradicate the deathwatch beetles. It is unlikely however, that this was because the chemicals were too fugitive, because samples taken from the surface 50 mm (2 in) of timber in 1935 still contained residues of sulphur and chlorine. [28] Fifty mm (2 in) is a remarkable depth of penetration for a preservative applied to the surface of oak and so the intended fumigant action was presumably partially successful, unless these residues came from the stove fumes. There are no indications, however, that these residues protected the timbers from subsequent attack, even though they must have been present in significant quantities to be detected by the analytical techniques then available. If we assume that the volatile ingredients had dispersed by 1950 then the other ingredients could only act as deterrents or be toxic by ingestion. This might simply leave an outer casing of timber which was unavailable to the beetles, while the core was vulnerable. Surface layer unavailability is of no consequence to the beetles if the emerging beetles do not have to bite their way out of the timber, and it can easily be demonstrated that beetles will emerge and re-enter wood via old flight holes, cracks and joints (Figures

Figure 5 Larvae may tunnel along the juvenile core of the timber where the natural durability is low.

Figure 6 The centre of the timber may eventually be hollowed by the beetles if an undetected heart rot was present when the timber was installed. See also Colour Plate 2.

Figure 7 An emergence hole, which was present when tissue paper was attached to the timber on 13 April 1997, was observed in use again on 14 April 1997. This experiment at the Tide Mill, Beaulieu, Hampshire, produced emergence holes in seven out of 80 papers used to cover areas of beetle damage.

Figure 8 A photograph taken with ultraviolet light shows a luminous trail left by a beetle which was lightly coated in a marker dust. The beetle re-entered the timber via a shake. (J Bustin). See also Colour Plate 3.

7 and 8). This behaviour helps to explain the frequently quoted observation that a few beetle holes do not necessarily indicate the extent of concealed damage (eg Hickin 1975).

Contact insecticides became readily available in the 1950s and these were used to treat the timbers on numerous occasions. Contact insecticides are absorbed through the cuticle of the insect, and do not need to be ingested before they will kill. They should therefore have overcome the disadvantages of the Lefroy/Heppells' formulations, particularly as beetles were now present in large quantities. The results however, were disappointing, as records from the east end of Truss 11 demonstrate. This truss end was spray-treated in 1952, 1954, 1956 and 1960 but by 1968 the number of beetles collected had increased to 887.

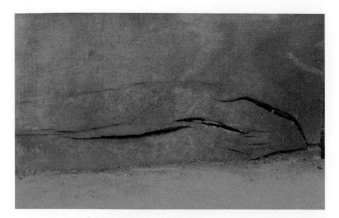

Figure 9 This photograph shows that the beetle re-entered the timber via an old flight hole. (J Bustin). See also Colour Plate 4.

The expected efficacy of contact insecticides for the control of deathwatch beetles came from Fisher's observations that the beetles laid their eggs on or in the surface of the timber, and that the newly hatched larva wandered extensively before burrowing. This led Hickin (1975) to conclude that eggs need not be laid on a suitable food source. Fisher's observations were made on beetles enclosed in laboratory containers, but beetles in their natural environment might behave rather differently. Figures 8 and 9 show trails left by live beetles which have been lightly coated with a luminous powder and released where they were found. In one the beetle retreated into a shake, while in the second it re-entered the timber via an old emergence hole. Recent research (Belmain *et al*, this volume) suggests that the female beetle re-enters the timber after mating in order to lay her eggs, and that oviposition will only occur on the surface of the timber if there is no other option. Spray treatment with contact insecticides will not therefore kill eggs and newly hatched larvae, the most vulnerable phases in the beetle's development. The ability to use old flight holes also diminishes contact with the insecticide. The only situation where surface treatments might prove effective is likely to be sapwood attack, because this is a surface layer phenomenon.

Smoke treatments seem to have been more effective, in that dead or moribund beetles could be shown by analysis to contain the contact insecticide (see Coleman, this volume). This increase in efficacy perhaps reflects a greater contact between beetles and a surface deposit of particulate insecticide, rather than one within the surface layer of the timber. Very little chemical has to be lost to the atmosphere in the latter situation before contact is significantly reduced between the insecticide and the insect, particularly if the beetle does not take wood into its mouth while biting an exit hole.

CONCLUSIONS

- The roof of Westminster Hall was dry in 1915 because of various repairs during the nineteenth century.
- The 1915–22 treatments did not eradicate the deathwatch beetles because they were at a low population density in inaccessible situations. Most of the timbers that were treated had been damaged a hundred or more years earlier.
- The roof remained dry during the interwar years, but the beetles persisted at a low population density.
- The beetle population increased dramatically in the years following bomb damage because the timbers were extensively wetted and remained damp for nearly 10 years.
- Subsequent surface treatments did not destroy the beetles, because the insect's natural behaviour protected vulnerable stages from contact with the insecticides.
- Smoke treatments were more effective but still did not eradicate the beetles.
- The beetle population eventually declined back to a low density when the roof dried. The beetles have not been eradicated, and eradication may not be possible with conventional insecticides.

ENDNOTES

1 See also the film *The Story of Westminster Hall*, made in 1923, and held by The National Film and Television Archive, British Film Institute, London

2 See also *1740–1837 Westminster Hall* (Work 11, 28/8) and *1816–1822 Report by John Soane* (Work 11, 28/10). (The Work 11 series of files are held at the Public Record Office, Kew.) File 28/8 also contains a *Report on the State of the Public Buildings at Westminster* dated 30 March 1816 by Soane, and both of Soane's reports indicate that the buildings were generally in a deplorable condition.

3 Report from Pearson to The Secretary, H M Office of Works, dated 7 May 1888 in *1887–1889 Westminster Hall Restoration, Part 2* (Work 11, 76, 1124/1).

4 Letter from Ridge to Hawks, in *1909–1914, Westminster Hall: examination and repair of roof* (Work 11, 196, 1534/41, Part 1).

5 Various reports, in *1909–1914. Westminster Hall: examination and repair of roof* (Work 11, 196, 1534/41, Part 1).

6 Identification of surface degradation which may have been caused by stove fumes in Westminster Hall, *Report of Work March/April 1983 in Westminster Hall DWB Treatment etc.* (File No. AA 008823/362, English Heritage Registry). Fumes from these stoves were known to have caused damage to the building: *1858–1859 Report by Professors W A Miller, and A S Taylor upon effect of coke fumes on the stonework, metalwork etc. of the building* (Work 11, 28/15).

7 Particularly *The Times* for 18 July 1913 and 20 August 1913 (doubting the extent of damage) and 26 January 1915 (the use of steel).

8 A copy in *1909–1914 Westminster Hall: examination and repair of roof* (Work 11, 1534/41, Part 1).

9 The suggestion was made in a letter from Dr A E Shipley (Christ's College, Cambridge) to Sir William Wedgwood Benn (First Commissioner of Works), in Work 11, 1534/41, Part 1.

10 The committee eventually consisted of the following: G J Gahan, British Museum (Natural History) G Marshall, Director of the Imperial Bureau of Entomology, J J Dobbie, Government Laboratory Professor Maxwell-Lefroy, Imperial College, London, Professor E Westergaard, Heriot-Watt University, Edinburgh, Dr A E Shipley, Christ's College, Cambridge, C Warburton (of Granchester, Cambridge).

The first meeting was held at 12 noon on Tuesday 25 November 1913 at H M Office of Works, Storey's Gate, London. (Work 11, 1534/41, Part 1).

11 Recorded in a letter from Shipley to Earle (Office of Works) dated 19 January 1914 (Work 11, 1534/41, Part 1).

12 Letter from Paget to Earle, 6 December 1913 (Work 11, 1534/41, Part 1).

13 In the letter from Paget to Earle, 6 December 1913 (Work 11, 1534/41, Part 1).

14 The disagreement became quite heated, for example the following is found in a letter dated 14 April 1914, from Westergaard to Baines.

'Many thanks for sending me Dr Dobbie's report which I will return. It is really so hopelessly wrong that I found considerable difficulty in drawing up a suitable reply.'

Shipley concludes in a letter to Earle dated 4 March 1914.

'I found our meeting yesterday rather disappointing, and the two chemists destroying each other's methods and then throwing over their own. In my opinion Lefroy, in spite of his emphatic manner and gloomy countenance, is the best man at the job' (Work 11, 196, 1534/41, Part 1).

15 See *Recommendations for the Treatment of the Roof Timbers to Prevent Further Damage by the Beetle* (Xestobium tessellatum), Maxwell-Lefroy's report to the Office of Works, in *1915–1923, Westminster Hall: examination and repair of roof* (Work 11, 217, 1534/41, Part 2).

16 Details of the treatment are contained in an unsigned or dated document entitled 'Wood-boring beetles. Notes upon their life history and measures adopted to exterminate them in Westminster Hall' in *DWB Treatment etc.* (File no. AA 008823/362, English Heritage Registry). Lefroy's agreement with the Hepple's mixture is in Lefroy to Baines, 22 October 1917. (Work 11, 217, 1534/41, Part 2).

17 In Work 11, 217, 1534/41, Part 2.

18 *1926–1949 Westminster Hall, inspection and treatment of roof timbers* (Work 11, 395, AE2534/41, Part 3).

19 Work 11, 395, AE2534/41, Part 3.

20 In *1940–1944, Reports of damage by enemy action* (Works 11, 399, AE2379/1, Part 1).

21 In *1941–1944 Bomb damage* (Work 11, 401, AE2509/40, Part 1).

22 In Work 11, 401, AE2509/40, Part 1.

23 In *1940–1954, Westminster Hall, wartime damage and repairs, supply of oak etc.* (Work 11, 439, AE 2534/43)

24 *1951–1954 Records of beetle attack*, Work 11, 597, AM50387/01D, Part 2.
1955–1959 Records of beetle attacks, Work 11, 598, AM50387/01D, Part 3.
1965 Records of beetle attack, AM50387/01D, Part 8, English Heritage, Registry.
1966 Records of beetle attack, AM50387/01D, Part 9, English Heritage, Registry.
1967 Records of beetle attack, AM50387/01D, Part 10, English Heritage, Registry.
1968 Records of beetle attack, AM50387/01D, Part 11, English Heritage, Registry.
1969 Records of beetle attack, AM50387/01D, Part 12, English Heritage, Registry.
1970 Records of beetle attack, AM50387/01D, Part 13, English Heritage, Registry.
1971 Records of beetle attack, AM50387/01D, Part 14, English Heritage, Registry.
1975 and 1977 in loose files, no titles, English Heritage Registry.
1980 records are in *Westminster Hall, DWB Treatment* at AA008823/362 English Heritage Registry.

25 In *1946–1968, Repair and Maintenance of roof* (Works 11, 596, AM50387/01D).

26 28 April 1966 a report from E C Harris to C G Wilson, Ministry of Public Building and Works. Westminster Hall, entitled *Examination of Insect-infected Roof Timbers*. Also a letter from C G Wilson to E C Harris in *1966 Records of beetle attack*, AM50387/01D, Part 9.

27 *Report of Deathwatch beetle survey* from Berry to Ashurst dated 7 April 1987 in *Westminster Hall DWB Treatment etc.* (File no. AA 008823/362, English Heritage Registry).

28 Surface chippings and borings were taken to a depth of 6 inches (150 mm) by F Cann during his 1935 inspection. These samples were tested for chlorine and sulphur by Dr K F Taylor of the Forest Product Research Laboratory. Residues were found in surface chippings and borings to a depth of 2 inches (50 mm). (Works 11, 395, AE2534/41, Part 3).

BIBLIOGRAPHY

Baines F, 1914 *Report to the First Commissioner of H. M. Works on the Condition of the Roof Timbers of Westminster Hall, with Suggestions for Maintaining the Stability of the Roof*, London, HMSO.

Becker G, 1942 Ökologishe und physiologische Untersuchungen über die holzzerstörenden Larven von *Anobium punctatum* Deg., in *Zeitschrift für Morphologie und Ökologie Tiere.* **39**, 98-151.

Blake E G, 1925 *Enemies of Timber: Dry Rot and the Death Watch Beetle*, Chapman and Hall, London.

Cartwright, K St G and Findlay W P K, 1946 *Decay of Timber and its Prevention*, London, HMSO.

Goulson D, Birch M C and Wyatt T O, 1993 Paternal investment in relation to size in the deathwatch beetle, *Xestobium rufovillosum* (Coleoptera: Anobiidae), and evidence for female selection for large males, in *Journal of Insect Behaviour* **6**:5, 539–547.

Harris E C, 1969 Assessment of insecticidal smokes for the control of wood-boring insects, *Record of the 1969 Annual Convention of the British Wood Preserving Association*, 5–23.

Hickin N E, 1975 *The Insect Factor in Wood Decay* (3rd edn), The Rentokil Library, London.

Maxwell Lefroy H, 1924 The treatment of the deathwatch beetle in timber roofs, in *Journal of the Royal Society of Arts*, **52**, 260–266.

Meyer O E, 1970 On adult weight, oviposition preference and adult longevity in *Anobium punctatum* (Col. Anobiidae), *Zeitschrift für angewandte Entomologie* Sounderdruck aus Bd., **66**, 103–112.

Munro J W, 1931 Insects injurious to timber, in *Journal of the British Wood Preserving Association*, **1**, 51–70.

Ridout B V, 2000 *Timber Decay in Buildings and its Treatment: The Conservation Approach*, Routledge, London.

Spiller D, 1948 An investigation into numbers of eggs laid by field collected *Anobium punctatum* De Geer, New Zealand, in *Journal of Science and Technology (section B)*, **30**: 3, 153–162.

Williams L H, 1983 Wood moisture levels affect *Xyletinus peltatus* infestations, in *Environmental Entomology*, **12**:1, 135–140.

AUTHOR BIOGRAPHY

Brian Ridout is a Director of Ridout Associates, consultants specializing in complex damp and decay investigations, expert witness work, scientific research and lecturing. He holds degrees from the Universities of Cambridge and London in entomology and mycology, and is a Fellow of the Institute of Wood Science.

Integrated pest management for the control of deathwatch beetles
Trapping

MONIQUE J SIMMONDS[1*], STEVEN R BELMAIN[2, 3] AND WALLY M BLANEY[2]

[1] The Jodrell Laboratory, The Royal Botanic Gardens, Kew, Richmond, Surrey TW9 3AB, UK;
Tel: +44 (0)20 8332 5328; Fax: +44 (0)20 8332 5340; email: M.Simmonds@rbgkew.org.uk, www.rbgkew.org

[2] Department of Biology, Birkbeck College, University of London, Malet Street, London WC1E 7HX, UK;
email: W.Blaney@bbk.ac.uk, www.bbk.ac.uk

[3] Natural Resources Institute, Central Avenue, Chatham Maritime, Kent ME4 4TB; Tel: +44 (0)1634 883 761; Fax: +44
(0)1634 883 567; email: S.R.Belmain@gre.ac.uk, www.nri.org

Abstract

Three different types of traps were used to monitor adult deathwatch beetles in historic buildings in England. The diversity of arthropods caught by unbaited sticky traps and insectocutor traps varied but the proportion of deathwatch beetles caught ranged from 30–40%. The numbers of beetles caught each week increased during the emergence season from May to July, as the temperature increased. Flying beetles were caught when the ambient temperature was above 17°C. White card sticky traps baited with ether and acetone extracts of *Donkiapora*-infected oak wood caught more beetles in a church than methanol or unbaited control traps. Acetone baited-sticky traps also trapped female beetles when placed in oak trees. The traps provided information about the diversity of arthropods found in roof spaces, including potential biological control agents such as predatory beetles and spiders. The potential use of sticky traps and ultra-violet-insectocutors in the monitoring and control of deathwatch beetles is discussed.

Key words

Deathwatch beetles, *Xestobium rufovillosum*, sticky traps, UV-insectocutor, monitoring, integrated pest management

INTRODUCTION

It is difficult, if not impossible, to estimate the damage done every year by the deathwatch beetle, *Xestobium rufovillosum* De Geer, to buildings in Europe. One major problem in estimating the damage, and in controlling the beetles, relates to the difficulties of identifying the areas in a building that are currently infested with beetles. The presence of round holes in timber made by emerging adult deathwatch beetles does not always indicate that the timber supports an active population of beetles. These holes could have been made by beetles emerging at any time in the last few centuries. Beetle infestations usually occur in timbers that have been damaged by fungi and/or water and are often found in parts of a building where the humidity is high, such as timbers adjacent to masonry works or gutters. Often infestations occur in parts of buildings that are difficult to reach and therefore difficult to treat. Traps have been used successfully to catch adult deathwatch beetles during the emergence season (Belmain *et al* 1999). These traps could form part of a monitoring strategy to establish the level and source of an infestation

* Author for correspondence.

of beetles in a building, allowing infested timbers to be localized, treated or replaced before infestations spread into other timbers within a building. This paper expands on the results presented in an earlier report (Belmain *et al* 1999) into whether different types of traps could be used to monitor or even control populations of beetles as part of an integrated pest management programme (IPM).

HISTORIC BUILDINGS USED IN THE STUDY

Winchester Cathedral

Winchester Cathedral was built on the site of an Old Minster which was destroyed in 1090 and a New Minster which was destroyed in 1110. The transepts of the present cathedral date back to 1107 when Bishop Walkelin started what is now known as the Cathedral. The building is situated on former water meadows and the Normans built the cathedral on a timber raft to prevent it from sinking into the peat bed. However, it has been subject to subsidence and has undergone major modification in every century. In the early twentieth century the timber raft was replaced with a concrete foundation. Two fires in the west end of the building required this area to be rebuilt and the nave walls were reinforced with iron cross-ties in the roof and external buttresses. Because of the situation of the cathedral, the crypt is regularly flooded and the building has chronic moisture problems. Ventilation of the roof has been increased but the humidity remains high at between 55–95%RH. Evidence of deathwatch beetle damage is very apparent in the timbers, including those replaced in the seventeenth to nineteenth centuries.

Salisbury Cathedral

Salisbury Cathedral was commissioned by Bishop Richard Poore in 1217. The construction started in 1220 and it was consecrated in 1258. The West Front was completed in 1265, the cloisters date from between 1263 and 1270 and the Chapter House was finished in 1284. Since construction the building has had relatively few modifications. However, the cathedral was neglected during the seventeenth century under Cromwell and the fabric of the church has deteriorated over the last two centuries due to poor maintenance. This has resulted in roof

Figure 1 Coloured sticky cards placed on the wall plate in the roof of Winchester Cathedral. See also Colour Plate 5.

timbers being exposed to moisture through lack of attention to the lead coverings. Currently the building is drying down, although the humidity levels still range from 50–80%RH. Deathwatch beetle infestations in the upper roof areas have required several central cross-ties and posts to be replaced due to structural weakening. However, beetles continue to be a problem in the upper roof.

Bishopstone Church

This thirteenth-century church situated by the River Ebble in Wiltshire is thought to be one of the most severely deathwatch beetle infested buildings in Britain. The church underwent major modification in the four-teenth century when the wooden ceiling of the chancel and south transept were replaced with stone vaulted ceilings. In the seventeenth century the sloping roof was placed on top of the previous flat roof. This modification resulted in a reduction in the ventilation within the building and an increase in humidity, which has most likely been associated with the deterioration of the timbers in the church roof. It appears that infestations of the beetles in the original thirteenth-century roof timbers have spread to the seventeenth-century pews and nine-teenth-century pulpit, choir stalls and other wooden structures in the church. The relative humidity in the winter months can be as high as 95%. Most of the structural deterioration of the timbers has been caused by beetles as there is very little evidence of fungal damage. There are signs that the timbers have been previously treated with tar-based pastes which might have contained lindane or DDT.

Kew Palace

The Dutch House, which is situated in the grounds of the Royal Botanic Gardens, Kew, and known as Kew Palace, was built in 1631. The design of the Palace with its gabled roof and valley gutters has caused moisture problems resulting in both fungal and beetle infestations. A major reconstruction of the roof area took place during the late twentieth century with the removal of all the plaster work and almost the entire western half of the roof timber was replaced with creosote-treated softwood. In 1997 the building underwent further major repairs due to dry rot in the bonding timbered brick work on the north side.

DIFFERENT TRAPS TESTED AND METHODS USED

A series of different types of traps were used to assess the populations of beetles in the buildings; these included coloured sticky cards and strips, and ultra-violet insectocutors (Belmain *et al* 1999). The traps were placed in areas thought to be infested with beetles. It was hoped that the traps would provide information about the diversity of arthropods in the roof area of the buildings, monitor the presence of biological control agents, such as predators or parasitoids, and evaluate whether the emer-gence of deathwatch beetle adults was influenced by environmental conditions. The effect of extracts made from oak wood decayed by the fungus *Donkioporia expansa* on the trapping efficiency of white card sticky traps was also assessed in Bishopstone Church and in oak trees at the Royal Botanic Gardens, Kew.

Sticky traps

Populations of agricultural and horticultural pests are frequently monitored using sticky traps. The efficiency of these types of traps has often been improved by including specific colours (Coli *et al* 1992, Vernon & Gillespie 1995) or chemicals (Sakuma & Fukami 1993). Although sticky traps have been used to monitor timber pests, very little research has been undertaken to evaluate whether they could be made more efficient by changing their colour or by including some form of chemical bait. Pheromone traps are available for the anobiid beetle, *Anobium punctatum,* using the female sex pheromone

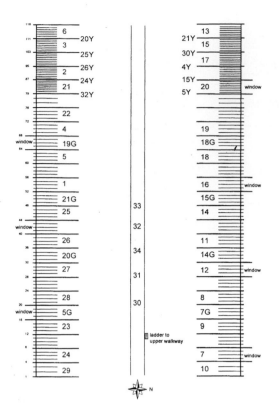

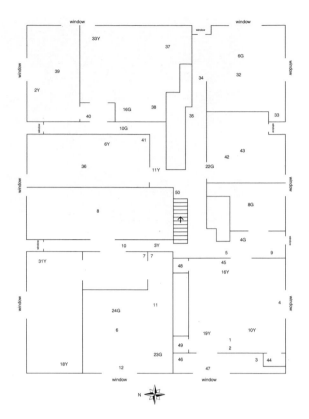

Figure 2 Location of sticky traps in Winchester Cathedral (drawing not to scale). Large numbers only refer to a group of four cards 100 mm x 150 mm/4 in x 6 in, one of each colour red, blue, yellow and white. Numbers with a 'Y' are yellow/brown strips and numbers with a 'G' are green strips hung from beams.

Figure 3 Location of sticky traps in Kew Palace roof (drawing not to scale). Numbers only refer to a group of four cards 100 mm x 150 mm/4 in x 6 in, one of each colour red, blue, yellow and white. Numbers with a 'Y' are yellow/brown strips and numbers with a 'G' are green strips hung from beams to the floor.

stegobinone (Birch & White 1988). The pheromone is the same as that produced by another anobiid beetle, the drugstore beetle, *Stegobium paniceum*, from which the pheromone was first identified (White & Birch 1987). Preliminary experiments by the group headed by Dr Birch at Oxford University have shown that although the death-watch beetle is thought to be related to these anobiids it does not respond to the sex pheromone stegobinone. They have been unable to demonstrate any behavioural response of deathwatch beetles to pheromones (pers comm).

Materials and methods
Red, yellow, blue or white paper cards 100 mm by 150 mm (4 in x 6 in) were covered with a glue called Tangle Trap®. Strips 50 mm wide and approximately 3 m (2 in x 3 yds) long were cut from rolls of green or yellow-brown wallpaper covered in Tangle Trap® and then rolled up. The Tangle Trap® had been heated in a water bath to 100°C to reduce viscosity and applied as thinly as possible to one side of the cards or strips with a 100 mm wide (4 in) paint brush. The sticky cards were placed on horizontal surfaces in groups of four, consisting of one card of each colour. The distance between the cards in a group and between groups varied due to the structure of the roof, but distances between cards in a group were never greater than 500 mm (1 foot 8 ins) (Fig 1). The position of each colour within a group was randomised.

In Winchester Cathedral, 136 numbered sticky card traps were placed along the wall plates on either side of

the nave roof, roughly in alternate bays. Some traps were placed along a gangway running along the centre of the roof at its peak (Fig 2). The traps were placed on 19 April 1995 and brought back to the laboratory on 12 July 1995.

In the attic rooms in Kew Palace, 120 sticky cards were placed on the floor, near walls, in window sills and along timber beams at head height (Fig 3). The traps were placed on 1 May 1995 and brought back to the laboratory on 12 July 1995.

In Salisbury Cathedral, 124 traps were placed out on 25 April 1995, mainly along the wall plates of the nave and in the north choir roof (Fig 4), they were brought back to the laboratory on 24 July 1995.

Within each building, thirty sticky strips were also hung in the same area as the cards. The sticky strips were hung in the buildings at the same time as the cards. In Kew Palace one or two strips were hung in each room, whereas at Winchester Cathedral and Salisbury Cathedral (Fig 5) they were hung along both sides of the nave roof. All sticky cards and strip traps were checked for insects weekly.

Ultra-violet (UV)- insectocutor

The efficacy of UV-insectocutors as a means of attracting and trapping insects is well researched (Roberts *et al* 1992, Mohan *et al* 1994, Frick & Tallamy 1996). However, their use in buildings to monitor or control timber pests is unreported and it was unknown if deathwatch beetles would be attracted to UV light. However, many

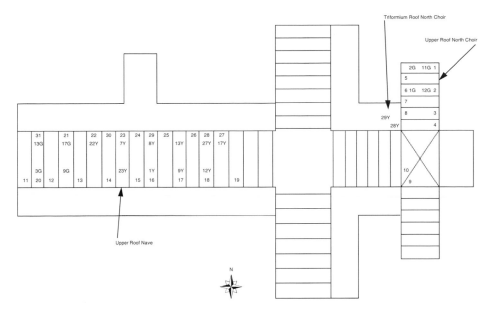

Figure 4 Location of sticky traps in Salisbury Cathedral roof. Numbers only refer to a group of four cards 100 mm x 150 mm/4 in x 6 in, one of each colour red, blue, yellow and white. Numbers with a 'Y' are yellow/brown strips and numbers with a 'G' are green strips hung from beams to the floor.

insects do orientate to UV light (Matthews & Matthews 1978) and it was possible that the beetles would show a similar response.

Materials and methods

The UV-insectocutor (NPW 80 Insectocutor, Pest West® Electronics Ltd.) was tested at Bishopstone Church because the building contained an active population of beetles and the roof area was small enough to be covered by the insectocutor (maximum coverage of 300 m² [359 square yards]). The insectocutor was placed in the windowless nave roof during the first week of April 1996 and remained there until July 1997. Insects were collected from the insectocutor on a weekly basis during April, May and June in 1996 and 1997.

Extract-baited sticky traps

Deathwatch beetles are often found in decayed oak wood, especially oak wood infected with the fungus *Donkiaporia expansa* (Brian Ridout pers comm). Laboratory experiments have shown that female deathwatch beetles select old oak in preference to new wood (Belmain 1998). However, it is not known whether beetles are attracted to decaying oak wood. If they are, then an extract from decaying oak wood incorporated into a trap as a bait might increase the efficiency of the traps. An experiment was set up to evaluate whether extracts from *Donkiaporia*-infected oak wood would influence the number of beetles caught on white card sticky traps in buildings. The white card was selected, as the earlier

Figure 5 An example of a series of sticky strips hanging down one side of the nave roof in Salisbury Cathedral.

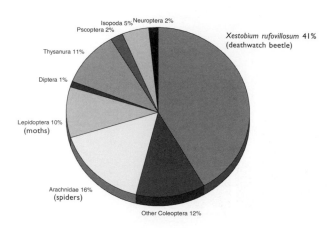

Figure 6 Total number (%) of arthropods, minus the group of arthropods classed as non-resident arthropods, caught on coloured sticky traps in Kew Palace, Winchester Cathedral and Salisbury Cathedral from April to July 1995.

sticky card trap experiment had shown that higher numbers of deathwatch beetles were caught on this colour of trap, compared to the red, blue or yellow coloured sticky traps.

We also took the opportunity to see if we could attract deathwatch beetles living in trees to these baited traps. Beetles are known to occur in old oak and willow trees but there is very little information about their ecology outside buildings. In this experiment we selected some mature oak trees growing in the Royal Botanic Gardens, Kew near Kew Palace to place the baited-sticky traps.

Materials and methods
Donkiaporia expansa-infected oak wood was supplied by Ridout Associates. The infected wood was macerated in a blender and 500 grams extracted in ether (500 ml) for 12 hours. The extract was decanted and the wood re-extracted in acetone (500 ml) for 12 hours, the solvent was decanted and the wood re-extracted in 80% methanol (500 ml) for 12 hours. Each extract was evaporated down to 50 ml and stored in sealed vials in a cold room (5° C). An aliquot (1 ml) of an extract was applied to an absorbent cotton strip (2 x 100 mm) that was then folded into a square (2 x 20 mm) and stuck to the centre of a

white card sticky trap, prepared as described above. Ten traps were prepared per extract; five traps were used for monitoring the deathwatch beetles in Bishopstone Church and five for monitoring the beetles in oak trees at the Royal Botanic Gardens, Kew. Ten unbaited control sticky traps were divided between the sites. The traps were placed in Bishopstone Church on 30 April 1999 and removed on 1 June 1999. Four traps (one per extract solvent and a control) were placed in the lower branches of five mature oak trees at Kew on 1 May 1999 and removed on 1 June 1999. The number of deathwatch beetles on each trap was recorded at the end of the exposure period.

RESULTS

Sticky coloured traps

In Winchester and Salisbury cathedrals and Kew Palace, beetles were predominately caught on the sticky traps over a six to seven week period from May to July 1995. The numbers of each species of arthropod caught on the traps in all three buildings from April and July 1995 were combined and divided into three different categories (Fig 6). These categories were based on the potential role of each species of arthropod in the ecology of the building. The three broad categories were:

- non-resident arthropods, such as bees or thrips, that are usually herbivorous and enter the building through windows and become trapped inside, but would find it difficult to survive inside a building
- over-wintering arthropods, such as many lacewings or moths
- detritus- and timber-eating resident arthropods, such as deathwatch beetles, firebrats, wood lice and carpet beetles as well as potential predators and parasitoids.

Analysis of the trap data revealed that 74% of the arthropods caught on the coloured sticky cards belonged to categories 2 and 3 and could be classed as 'resident arthropods'. Non-resident arthropods trapped on the

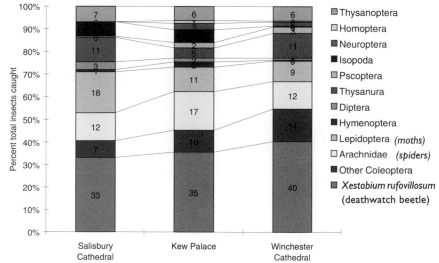

Figure 7 Comparison of the total number (%) of arthropods caught on coloured cards at Winchester Cathedral, Kew Palace and Salisbury Cathedral between April and July 1995. See also Colour Plate 6.

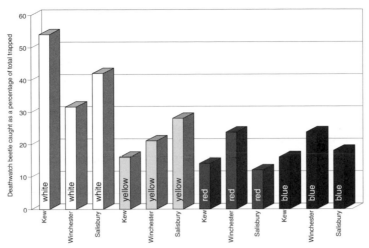

Figure 8 Percentage of deathwatch beetles caught on the different colours of traps at Kew Palace, Winchester Cathedral and Salisbury Cathedral in 1995.

cards were removed from the dataset to present a more meaningful description of the diversity of arthropods within a building (Fig 7). On the sticky strip traps the proportion of arthropods caught at the three sites was similar to that on the coloured cards, although only flying insects were caught. Resident arthropods comprised 52% of the insects trapped on the sticky strips.

In all three buildings deathwatch beetles made up the majority of insects caught in the roof areas during the trapping period using both coloured sticky cards and strips. On the coloured sticky cards, deathwatch beetles comprised 33% to 40% of the insects trapped in the buildings. The number of arthropods caught was also shown to be positively related to the maximum air temperature in the buildings during the period April to July.

The resident arthropods included some that prey on deathwatch beetles. For example, spiders (Arachnids) comprised 16% of the arthropods caught on coloured cards in the buildings. The other predator of the death-

watch beetle was the Clerid beetle, *Korynetes caeruleus*, which was caught on cards and strips at all three sites, although in small numbers (< 1% of total catch). The majority of other arthropods in the 'resident' category were detritus and wood feeders, ie Thysanura, Pscoptera, Isopoda and other wood-eating coleopterans. These arthropods could contribute to the overall deterioration of the building structure.

Analysis of the distribution of the deathwatch beetles caught on the different coloured cards demonstrated that white traps caught more beetles than yellow, red or blue traps ($c^2 > 11.2$, P < 0.001) (Fig 8). White cards were not only found to be more likely to have beetles on them, but also to have a greater number of beetles on them than cards of any other colour (Kruskal–Wallis, $c^2 = 56.771$, P < 0.001) (Fig 9).

The position of cards near and away from windows enabled us to investigate whether light could have influenced the number of beetles caught on the cards. The

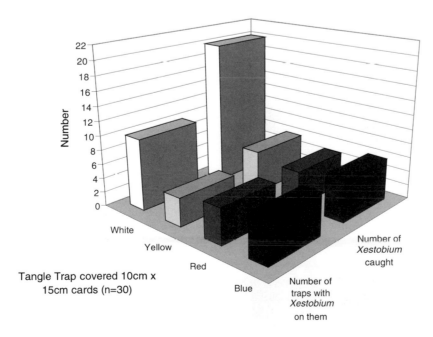

Figure 9 Distribution of deathwatch beetles caught on the different coloured traps and the proportion of traps that had caught beetles in 1995. The horizontal axis shows Tangle Trap™-covered 100 mm x 150 mm cards (n=30).

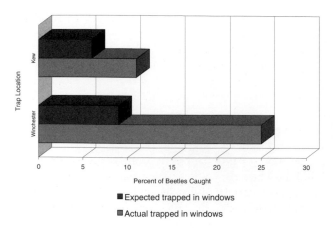

Figure 10 Number (%) of deathwatch beetles caught on sticky traps near windows in Kew Palace and Winchester Cathedral. The expected values were determined by assuming that the beetles would have been distributed equally between traps in the light and those in the dark.

results varied among the buildings but overall, more deathwatch beetles were caught on traps in the light than in the dark ($c^2 > 15.5$, P < 0.01) (Fig 10). The number of windows in the roof area of Kew Palace ensured that the contrast between light and dark was less in this

building than in the others, yet nevertheless more beetles were caught in traps near the windows than in other areas of the attic roof.

The sex of the beetles caught on the traps was determined through dissection after the traps were removed from the buildings. It was found that females outnumbered males by 2 to 1. There was no difference in this 2:1 sex ratio between traps in the light as opposed to those in the dark nor on traps of any particular colour (Mann-Whitney U-tests and Kruskal-Wallis, p > 0.5).

Ultra-violet (UV) insectocutor

Weekly variations occurred in the proportion of arthropods caught in the insectocutor (Figs 11a and 11b). Deathwatch beetles were caught in the UV-insectocutor from May to July and comprised 31% of all insects caught by the insectocutor during the collection periods. Most of the other insects caught were potentially detrimental to buildings, as they feed upon wood and decaying material. However, a few predatory beetles, *Korynetes caeruleus*, were caught. No deathwatch beetles were caught in the insectocutor until the weekly maximum temperature in the roof had exceeded 17°C and the

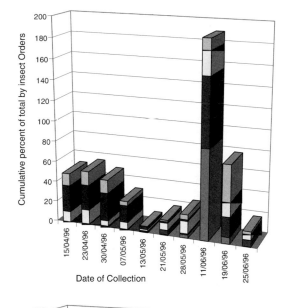

(a)

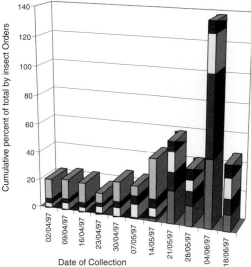

(b)

Figure 11 Number (%) of insects caught by the UV-insectocutor in the roof of Bishopstone Church from April to July in (a) 1996 and (b) 1997.

Table 1 Number of deathwatch beetles caught in Bishopstone Church on the white card sticky traps baited with extracts of decayed oak wood (five traps per extract).

Number of deathwatch beetles caught on the traps

Beetles	Control[a]	Ether	Acetone	Methanol
Female	10	34	59	13
Male	3	24	6	5
Total	13	58	65	18

[a] Control traps were unbaited with untreated absorbent cotton strips (2 x 100 mm) folded into a 2 x 20 mm square and placed in the middle of the sticky trap.

number of deathwatch beetles caught in the insectocutor remained relatively low until ambient temperatures were in the range of 24°C to 27°C.

Baited-sticky card traps

The white card sticky traps baited with the ether and acetone extracts caught a higher number of deathwatch beetles than the methanol-baited traps or unbaited control traps at Bishopstone Church (Table 1). The traps baited with the acetone extract caught more female beetles than any of the other traps. None of the traps caught any predatory spiders or clerid beetles.

Five female deathwatch beetles were caught on two of the acetone-baited white card sticky traps in oak trees at Kew. None of the other sticky traps in the oak trees caught any deathwatch beetles.

DISCUSSION

The unbaited sticky traps and insectocutor have been shown to be effective at monitoring adult deathwatch beetles and they have also provided ecological data on the types of arthropods present in roof habitats, including insects naturally dwelling in the roof as well as immigrants from outside the building. The presence of immigrant insects indicates that the internal areas under the roof are exposed to the external environment in ways that insects can penetrate. Differences in this porosity of the roof areas of the buildings might be reflected by the proportions of non-resident insects caught by the traps. The sticky strips

and insectocutor caught only flying insects, whereas the coloured sticky cards caught mainly walking arthropods including a high proportion of non-resident insects. Despite the variations in the numbers and types of arthropods caught by the different traps, the proportions of deathwatch beetles caught were similar at between 30% to 40% (Table 2). The exact levels of beetle infestations in the four buildings are unknown but observations of the amount of deathwatch beetle damage would suggest that the populations of deathwatch beetles in the buildings differed, with Bishopstone Church showing the greatest amount of damage and Kew Palace the least. Why, therefore, did the traps catch similar proportions of beetles in the different buildings?

The percentages of different species of arthropods caught by the traps may reflect the diversity of arthropods of the roof habitats, but the percentages caught are also a reflection of the efficiency of the trapping methods used. Habitat diversity is perhaps more clearly presented through the number of species caught as opposed to a percentage of the total population. Using the species gathered from the trapping methods at the four buildings, Bishopstone Church had the most diverse roof ecology, and Kew Palace had the least diverse habitat (Table 3). A trapping method that caught a large number of different species would be an appropriate method to evaluate the diversity of a habitat (Hagstrum *et al* 1990, Oakland 1996), whereas if a trap is to be used to monitor or control a specific insect then the relative proportions of species caught, other than the pest insect, should be low. Therefore data collected with a specialised and efficient collection method in a building with fewer species could resemble data collected with a less specialised and inefficient method in a building with a higher number of species (Hagstrum *et al* 1990). We need to undertake more experiments to assess which category the traps we used fall into.

The similar proportion of deathwatch beetles caught among the building and trapping methods may be a coincidence or it might reflect some form of relationship that exists between habitat diversity, building porosity and the degree of deathwatch beetle infestation in a

Table 2 Comparison among trapping methods and the historical buildings in the percent of total catch comprising deathwatch beetle adults

Building roof	Collection method	Deathwatch beetle %
Kew Palace	coloured sticky cards – 1995	35
Winchester Cathedral	coloured sticky cards – 1995	40
Salisbury Cathedral	coloured sticky cards – 1995	33
Kew Palace	sticky strips – 1995	34
Winchester Cathedral	sticky strips – 1995	38
Salisbury Cathedral	sticky strips – 1995	32
Bishopstone Church	insectocutor – 1996	30
Bishopstone Church	insectocutor – 1997	35

Table 3 Number of arthropod species caught by the different traps in the four buildings.

Arthropod type	Kew Palace 1995	Winchester Cathedral 1995	Bishopstone Church 1996	Salisbury Cathedral 1995
Resident	13	28	25	21
Over-wintering	8	15	33	12
Non-resident	17	14	6	20
Total number of species	38	57	64	53

roof (Holyoak *et al* 1997) that cannot be reliably explained by our experiments. Our ability to trap some beetles in oak trees indicates that these trees might contain populations of beetles that could infest buildings. However, if this is the case it is very surprising that there is little information in the literature about the occurrence of populations of deathwatch beetles in woodlands containing oak or willow trees or in the gardens or cemeteries associated with deathwatch-infested houses or churches, respectively.

Regardless of habitat diversity and trapping efficacy, we have shown that deathwatch beetles comprised the largest number of insects caught in Winchester and Salisbury cathedrals, Kew Palace and Bishopstone Church, in proportions surprisingly similar in each of them. These data suggest that the biodeterioration in these buildings can, at least in part, be attributed to the presence of active populations of deathwatch beetles.

Proof that deathwatch beetles can fly, coupled with evidence that the roof structures allow the entry of insects, would suggest that infestation of a building is possible centuries after construction, so that there is always the risk that a building may become re-infested. Thus traps could be used to monitor the presence of deathwatch beetles in treated as well as untreated roofs as part of a prevention, as well as a control, strategy. In fact, the traps could play an important role in monitoring the success of a remedial control programme.

Observations have shown that wild populations of beetles start to emerge at the beginning of April (Belmain 1998), a month before the insects were caught in the traps. This delay in catching beetles on the traps could be due to the effect of temperature on the behaviour of the insects. The low temperatures in April might have been below the threshold required for flight (Holyoak *et al* 1997). Research on the flight behaviour of other coleopterans showed that the number of insects caught in a light trap was often positively correlated to minimum temperature (Gupta *et al* 1990). Our results suggest that there could be a similar relationship between the number of deathwatch beetle trapped in the UV-insectocutor and maximum temperature. However, whether the relationship is causal needs further investigations. For example, our data do not enable us to distinguish whether an increase in temperature caused more beetles to emerge or to increase the activity of the beetles that had already emerged and thus the likelihood that they might be caught on a trap. It is likely that both factors contributed to the trap catches, although the numbers caught were ultimately linked to the emergence period, from April to June, as further increases in temperature during June and July did not result in further deathwatch beetles being caught.

However, other factors associated with the sex or physiological state of the insect may also contribute to the temporal distribution of beetles trapped. More females were caught in the insectocutor than males. Assuming a 1:1 sex ratio at emergence (Belmain 1998) the 2:1 ratio of females caught by the traps could imply that female beetles were more responsive to visual stimulation or were more active and flew more readily than males. However, this difference in the sex ratio caught could be due to the short lifespan of males. If males had emerged at the same time as females in April then a high proportion of them might have died before the ambient temperatures were high enough to induce flight. Further research would be required to determine whether flight behaviour differs between the sexes and whether this could explain why more females than males were caught.

The position of sticky traps in a building, as well as their colour, could influence their efficiency. The results showed that more insects were caught in traps near windows. However, this effect could be due to the relative apparency of the traps as traps in darker areas may not be as visible to the beetles. The uneven distribution of trapped insects between the light and dark areas could also result from environmental effects, such as differences in infestation levels or differences in the ambient temperature within the building. For example, more deathwatch beetles may have been emerging from timbers near windows as opposed to elsewhere in the building. Although the beetles appeared to differentiate between the coloured sticky traps, we have no evidence of spectral sensitivity. The white cards may have been more attractive because they presented a stronger contrast to the surrounding timber (Conlon & Bell 1991, Hughes 1992), or appear brighter than the other coloured cards (Kostal 1991) because they reflect more light than the red, yellow or blue cards.

Others have shown that trap efficacy can be enhanced by, for example, the addition of chemical baits (Finch & Skinner 1982, Harris & Miller 1988, Bursell 1990). Although Birch and his group at Oxford were not able to identify a pheromone that could be added to a trap to attract beetles, our results indicate that there may be other compounds from their preferred timber, oak, that could be used to bait traps. Our preliminary experiment with baited traps indicates that the addition of extracts from fungal decayed oaks could increase the efficiency of white card sticky traps. The active compounds in the extracts are unknown but appear to be apolar as the trap baited with apolar extracts (ether and acetone) caught more deathwatch beetles than those baited with the more polar extract (80% methanol). Further research should ideally be aimed at determining the identity of the active ingredient(s) in the extracts and determining the best way to incorporate the active ingredient into a trap designed to predominately trap deathwatch beetles.

There might be some environmental restraints in using baited or un-baited UV-insectocutors or sticky traps. The present results indicate that the effectiveness of insectocutors to monitor deathwatch beetles might be limited by low ambient temperatures. In a very cold spring or summer, low numbers of adults caught could lead to an inaccurate assessment of the level of infestations in a building. It is envisaged, however, that insectocutors could prove to be effective in reducing insect populations in buildings, especially if coupled with a heating system that could raise the ambient temperature above the threshold temperature required for insect flight early in the deathwatch beetle emergence season. Many buildings have heating systems and it may be possible to increase the temperature of the roof during the emergence season so that a higher proportion of the beetles could be caught by the insectocutor. The relative effectiveness and the cost of heating the building would have to be weighed against the cost of insecticidal treatment, timber replacement and/or other remedial maintenance required because of insect damage.

The use of an insectocutor in a removal trapping programme would justify further attention as a possible alternative to the broadcast application of chemicals for pest control. Removal trapping has been shown to be an effective method for the control of flies as vectors of disease (Day & Sjogren 1994). The insectocutor is a relatively simple method to implement, requires little maintenance and should be safe to use. Great care must be taken to ensure that the mesh size is small enough to exclude bats. Further investigations are underway to see if the efficiency of the insectocutor could be improved by the addition of the ether- or acetone-based baits from the decayed oak wood to the base of the insectocutor.

The diversity of insects trapped in the buildings indicates that there might be a role for biological control agents in the control of deathwatch beetles. Spiders comprised a large percentage of the arthropods caught on the sticky cards and the species caught on the traps have been observed to feed upon adult deathwatch beetles. However, the number caught of the predatory beetle, *Korynetes caeruleus,* was low, and its impact upon the deathwatch beetle remains obscure (Maxwell-Lefroy 1924, Fisher 1937, Hickin 1963, Birch & Menendez 1991). It is not known whether these predators could control beetles but they might modulate beetle populations. The low numbers of these predators caught on the traps might not reflect their density in the habitats but the low efficiency of the traps in catching them. The effectiveness of these predators is unknown and more work would need to be undertaken if they were to form part of an IPM programme. It is possible that the introduction of *Korynetes caeruleus* into deathwatch beetle-infested buildings could lead to a possible biocontrol solution as the predators attack deathwatch beetle larvae deep within timber (Belmain 1998). However, there are no published data to indicate their potential and many of the pesticides used to treat deathwatch beetle infestation could be more active at decreasing the populations of predators than beetles.

It was hoped that this study would identify other potential biocontrol organisms such as parasitoids, that lay their eggs in beetles and kill them, in addition to the predatory spiders and Clerid beetles. The absence of other biocontrol organisms from the survey results could be due to several reasons, eg:

- there are no other biological control organisms of the deathwatch beetle
- other biological control organisms occur in forests or in other building habitats not investigated

- other biological control agents were present but were not discovered by the methods implemented.

Although the types of traps used in this study may have overlooked potential control organisms, evidence from all the buildings and collection methods would suggest that if other biological control organisms against the deathwatch beetle exist, these organisms are neither widely prevalent in buildings attacked by deathwatch beetles nor found in significant numbers to have any substantial impact upon deathwatch beetle populations. Investigation into forest habitats, where the deathwatch beetle lives in dead timber, may prove more fruitful in identifying other biological control organisms. For example, potential microbiological control organisms have yet to be investigated.

Summary

- Adult deathwatch beetle were caught by either sticky traps or UV-insectocutors from May until July.
- Deathwatch beetles can fly and thus have the potential to migrate between buildings and from woods and forests into buildings.
- Deathwatch beetles were trapped in oak trees using sticky traps baited with acetone extracts from fungal-decayed oak wood.
- Sticky traps and UV-insectocutors can be used to monitor deathwatch beetles in buildings.
- White, highly reflective traps, catch more insects than other coloured traps.
- More beetles were caught on traps placed near sources of light than in the dark.
- White card sticky traps baited with an acetone extract from fungal-decayed oak wood attracted five times more beetles, especially females, than an unbaited control trap.
- Ambient temperatures influence the flight behaviour of beetles and the ambient temperature needs to be at least 17°C before the beetles will fly.
- If the temperature of an area infested with beetles could be raised above 17°C during the emergence season then UV-insectocutors could be used for both monitoring and controlling beetles.
- UV-insectocutors could be used as part of an integrated pest management (IPM) programme which included natural predators as the traps did not kill many predatory spiders, although they did catch a few predatory clerid beetles.
- If compounds that attracted beetles could be identified then there is the potential of using baited sticky traps or UV-insectocutors to monitor and control deathwatch beetles.

Overall, traps could play an important part of an IPM. They could be used as a non-environmentally harmful method to monitor and assess the populations of deathwatch beetles in buildings. However, there might be risks in using sticky traps and UV-insectocutors in areas frequented by bats and nesting birds. These risks could beminimizsed by improving the design of traps.

BIBLIOGRAPHY

Belmain S R, 1998 *The Biology of the Deathwatch Beetle* Xestobium rufvillosum *De Geer (Coleoptera: Anobiidae),* unpublished PhD thesis, University of London.

Belmain S R, Simmonds M S J and Blaney W M, 1999 Deathwatch beetle, *Xestobium rufovillosum,* in historical buildings: monitoring the pest and its predators, in *Entomologia Experimentalis et Applicata,* **93**, 97–104.

Birch M C and Menendez G, 1991 Knocking on wood for a mate, in *New Scientist,* 6 July, 42–44.

Birch M C and White P R, 1988 Responses of flying male *Anobium punctatum* (Coleoptera: Anobiidae) to female sex pheromone in a wind tunnel, in *Journal of Insect Behaviour,* **1**:1, 111–115.

Bursell E, 1990 The effect of host odour on the landing responses of tsetse flies (*Glossina morsitans morsitans*) in a wind tunnel with and without visual targets, in *Physiological Entomology,* **15**, 369–376.

Coli W M, Hollingsworth C S and Maier C T, 1992 Traps for monitoring pear thrips (Thysanoptera, Thripidae) in maple stands and apple orchards, in *Journal of Economic Entomology,* **85**:6, 2258–2262.

Conlon D and Bell W J, 1991 The use of visual information by house flies, *Musca domestica* (Diptera, Muscidae), foraging in resource patches, in *Journal of Comparative Physiology A Sensory Neural and Behavioural Physiology,* **168**:3, 365–371.

Day J F and Sjogren R D, 1994 Vector control by removal trapping, in *American Journal of Tropical Medicine and Hygiene,* **50**:6, 126–133.

Finch S and Skinner G, 1982 Trapping cabbage root flies in traps baited with plant extracts and with natural and synthetic isothiocyanates, in *Entomologia Experimentalis et Applicata,* **31**, 133–139.

Fisher R C, 1937 Studies of the biology of the deathwatch beetle, *Xestobium rufovillosum* De Geer: Part I, a summary of the past work and a brief account of the developmental stages, in *Annals of Applied Biology,* **24**, 600–613,

Frick T B and Tallamy D, 1996 Density and diversity of nontarget insects killed by suburban electric insect traps, in *Entomological News,* **107**:2, 77–82.

Gupta R C, Kundu H L and Thukral A K, 1990 Flight activity in some photopositive Coleopterans in relation to temperature, in *Journal of Environmental Biology,* **11**:4, 405–412.

Hagstrum D W, Flinn P W, Subramanyam B, Keever D W and Cuperus G W, 1990 Interpretation of trap catch for detection and estimation of stored-product insect populations, in *Journal of the Kansas Entomological Society,* **63**:4, 500–505.

Harris M O and Miller J R, 1988 Host acceptance behaviour in an herbivorous fly, *Delia antiqua,* in *Journal of Insect Physiology,* **34**, 179–190.

Hickin N E, 1963 *The Insect Factor in Wood Decay,* London, Hutchinson.

Holyoak M, Jarosik V and Novak I, 1997 Weather-induced changes in moth activity bias measurement of long-term population dynamics from light trap samples, in *Entomologia Experimentalis et Applicata,* **83**:3, 329–335.

Hughes R N, 1992 Effect of substrate brightness difference in isopod (*Porcellio scaber*) turning and turn alternation, in *Behavioural Processes,* **27**:2, 95–100.

Kostal V, 1991 The effect of colour of the substrate on the landing and oviposition behaviour of the cabbage root fly, in *Entomologia Experimentalis et Applicata,* **59**:2, 189–196.

Matthews R W and Matthews J R, 1978 *Insect Behaviour,* New York: John Wiley and Sons, Inc.

Maxwell-Lefroy H, 1924 The treatment of the deathwatch beetle in timber roofs, in *Journal of the Royal Society of Arts,* **72**, 260–270.

Mohan S, Gopalan M, Sundarababu P C and Sreenarayanan V V, 1994 Practical studies on the use of light traps and bait traps in the management of *Rhyzopertha dominica* (F) in rice warehouses, in *International Journal of Pest Management,* **40**:2, 148–152.

Oakland B, 1996 A comparison of 3 methods of trapping saprophytic beetles, in *European Journal of Entomology,* **93**:2, 195–209.

Roberts A E, Syms P R and Goodman L J, 1992 Intensity and spectral emission as factors affecting the efficacy of an insect electrocutor trap towards the housefly, in *Entomologia Experimentalis et Applicata,* **64**:3, 259–268.

Sakuma M and Fukami H, 1993 Aggregation arrestant pheromone of the German cockroach, *Blattella germanica* (L) (Dictyoptera, Blattellidae) – isolation and structure elucidation of Blattellastonoside A and Blattellastonoside B, in *Chemical Ecology,* **19**:11, 2521–2541.

Vernon R S and Gillespie D R 1995 Influence of trap shape, size, and background colour on captures of *Frankiliniella occidentalis* (Thysanoptera, Thripidae) in a cucumber greenhouse, in *Journal of Economic Entomology,* **88**:2, 288–293.

White P R and Birch M C, 1987 Female sex-pheromone of the common furniture beetle Anobium punctatum (Coleoptera, Anobiidae) - extraction, identification, and bioassays, in *Journal of Chemical Ecology,* **13**:7, 1695–1706.

ACKNOWLEDGEMENTS

This project was sponsored by the European Commission as part of the Woodcare research programme. We thank our collaborators; Brian and Elizabeth Ridout of Ridout Associates, John Fidler of English Heritage, Dervilla Donnelly and Hubert Fuller of University College Dublin, and Petra Esser and Jan de Jong of TNO in the Netherlands as well as the staff at Winchester and Salisbury cathedrals, Bishopstone Church and Kew Palace for allowing us access to the buildings.

BIOGRAPHIES

Professor M S J Simmonds is Head of Biological Interactions at the Royal Botanic Gardens, Kew, a group that studies the role of plant- and fungal-derived compounds on insect behaviour. She assisted Professor Blaney in the coordination of the EU-funded Woodcare project and is currently working with Drs Brian and Elizabeth Ridout of Ridout Associates on the development of traps to monitor the beetles.

Dr S R Belmain is a Senior Research Scientist at the Natural Resource Institute, Greenwich University working on the control of insect pests of stored products. He was the Research Associate employed by Birkbeck College on the EU-funded Woodcare project. Part of the work presented in this paper contributed to his doctoral thesis (Belmain 1998).

Professor W M Blaney is an Emeritus Professor at the Department of Biology, Birkbeck College, University of London his expertise is in insect physiology. He was the Birkbeck co-ordinator for the EU-funded Woodcare project that supported the work presented in this paper.

Plate 1 *Emerging* Xestobium rufovillosum *adult. See Figure 1, Page 7.*

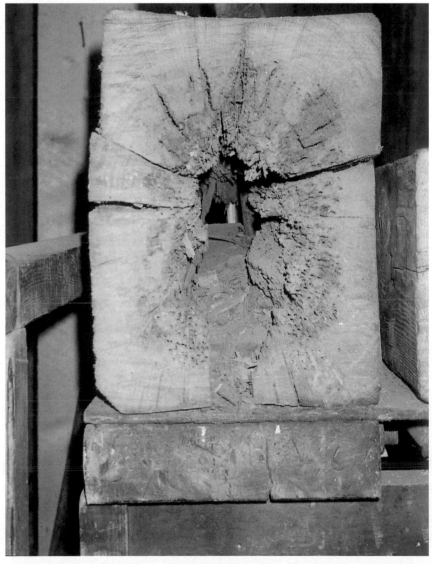

Plate 2 *The centre of the timber may eventually be hollowed by the beetles if an undetected heart rot was present when the timber was installed. See also Colour Plate 3. See Figure 6, page 37.*

Plate 3 *A photograph taken with ultraviolet light shows a luminous trail left by a beetle which was lightly coated in a marker dust. The beetle re-entered the timber via a shake. (J Bustin). See also Figure 8, page 37.*

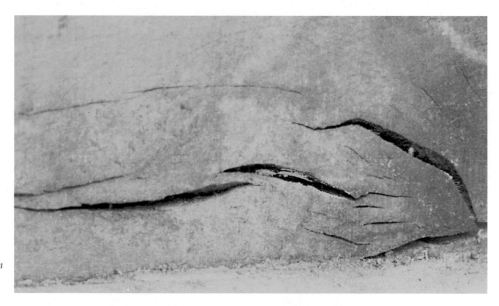

Plate 4 *This photograph shows that the beetle re-entered the timber via an old flight hole. (J Bustin). See also Figure 9, page 38*

Plate 5 *Coloured sticky cards placed on the wall plate in the roof of Winchester Cathedral. See also Figure 1, page 41.*

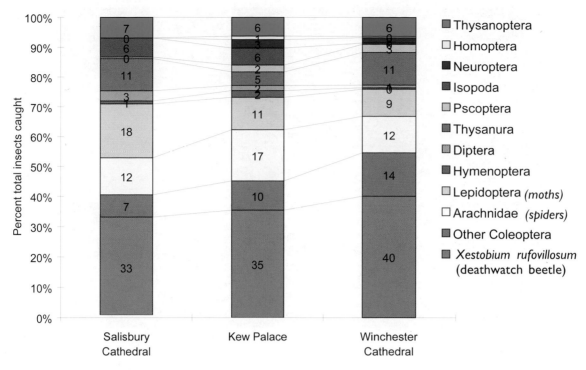

Plate 6 *Comparison of the total number (%) of arthropods caught on coloured cards at Winchester Cathedral, Kew Palace and Salisbury Cathedral between April and July 1995. See also Figure 7, page 44*

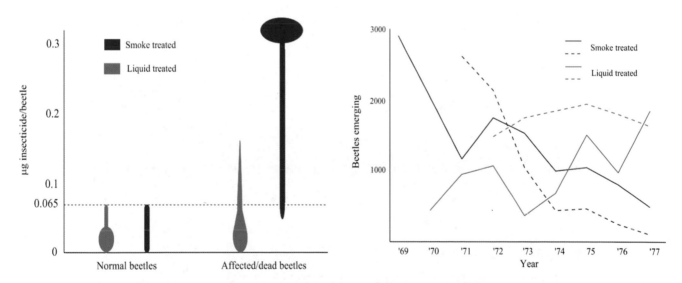

Plate 7 *Pick-up of insecticide vs conditions of beetles. See also Colour Figure 1, page 52.*

Plate 8 *Egg-laying 1969–77, shown by years against 1000s. See also Figure 3, page 53.*

Plate 9 *Spreading basidiocarp of* Donkioporia *on oak beam in roof at Stoneleigh Abbey, Warwickshire, UK (April 1997). Bar = 100 mm. See also Figure 6, page 60.*

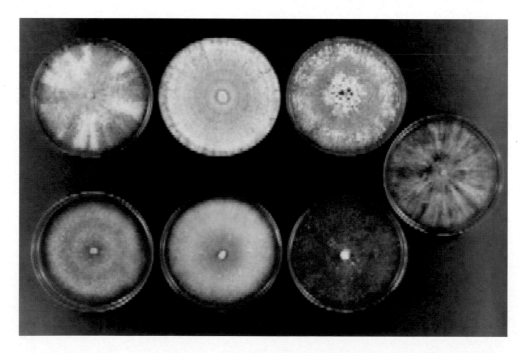

Plate 10 *Variation in colony morphologies of two-week-old* Donkioporia *isolates on potato dextrose agar. From top left, clockwise: 236.91, 299.93, 30747, 31565, 30819, 178E and 31593. 90 mm Petri dish. See also Figure 8, page 62.*

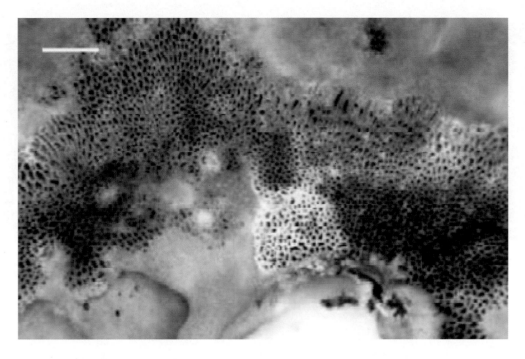

Plate 11 *Culture of* Donkioporia *(Isolate 30747) showing poroid formation at the edge of a Petri dish. Bar = 10 mm. See Figure 10, page 63.*

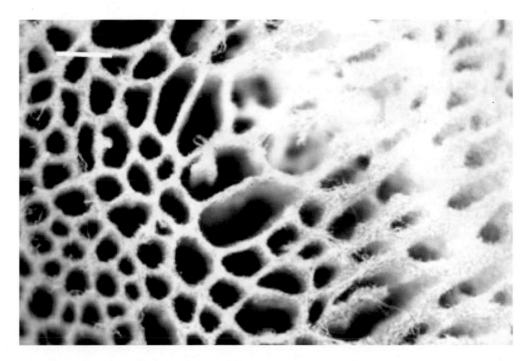

Plate 12 In vitro *differentiation of dissepiments varying from angular to round in shape, of* Donkioporia *on potato dextrose agar at 25°C. Bar = 1 mm. See Figure 11, page 63.*

Plate 13 *A photograph of a rendered urban building of indeterminate age and construction on the corner plot of a block of buildings of varying dates and styles. See also Figure 1, page 101.*

Plate 14 *The thermographic overlay of the building in Figure 1 identifies a number of interesting points. See also Figure 2, page 101.*

Plate 15 *The thermal image of this church ceiling identifies the secondary timbers concealed behind the plaster. More significantly, a damp patch (arrowed) indicates a leak in the roof that could not be accessed from above the ceiling. Had this not been identified, the persistent wetting would have encouraged fungal and insect attack. See Figure 9, page 104.*

Plate 16 *The rear elevation of a property, constructed of brick to ground floor and cement-based roughcast to first floor. The date of the building and materials of the first floor are indeterminate. See also Figure 3, page 102.*

Plate 17 *The same elevation as in Figure 3 with a thermographic image superimposed on the photograph. Immediately, and with no opening up or special access, the method of construction, the exact location of the individual elements and other useful information are clearly defined. See also Figure 4, page 102.*

Plate 18 *An internal wall of Kenwood House, London. There is no indication of the structure behind the fine decorative plaster. See also Figure 7, page 103.*

Plate 19 *The thermographic image superimposed on the photograph clearly shows the structure, and its exact location. See also Figure 8, page 103.*

The chemical control of deathwatch beetle

G R COLEMAN

Remedial Technical Services, 14 Mill Rise, Bowton, nr Gillingham, Dorset SP8 5DH, UK;
Tel: +44 (0)1747 840715; www.mill.rise.freeserve.co.uk

Abstract

Nearly twenty years of research involving the assessment of over 60,000 beetles showed that spray treatments had no lasting effect on beetle populations. Pyrotechnic smoke formulations were far more effective, but suitable smokes are not commercially available at the present time. Smokes are non-targeted, but may be useful to deplete beetle numbers in buildings where there is heavy infestation.

Key words

Deathwatch beetle, spray treatment, smoke treatment, insecticides

INTRODUCTION

Throughout the 1970s a team of scientists from the Building Research Establishment (BRE) and Princes Risborough Laboratories (PRL) continued and expanded work begun by Ernest Harris in the 1960s on the practical chemical control of deathwatch beetle. Over 60,000 beetles were collected and assessed for the effects of insecticides during the course of the research. The efficacy of different types of chemical treatments on the numbers of beetles emerging was assessed. These treatments are compared in this report with 'natural' methods of control.

There is a general trend away from the use of chemicals, but this report shows that in targeted situations chemicals can be beneficial in the control of deathwatch beetle.

'NATURAL' APPROACHES TO BEETLE CONTROL AND THEIR LIMITATIONS

Dry out the wood

Wood-boring insects cannot survive in wood at less than 9–10% moisture content, and, theoretically, if wood is allowed to dry down to these levels, then the insects will die. Unfortunately maintaining wood at these moisture contents is not often possible because wood is hygroscopic and absorbs moisture from the air. More humid air leads to greater water absorption.

The practicalities of this theory are complex. Table 1 shows that a relative humidity (%RH) of 45% is required to maintain wood at an equilibrium moisture content of 10%. This seems impractical when conditions in a typical church may be $6^0C/88\%$ (the equivalent of an equilibrium moisture content of 20%).

The temperature will be less in the ends of the timbers built into the wall, and the humidity will be higher there. The moisture content of the ends of timber beams embedded in cold walls will be high enough to allow survival of wood-boring larvae.

Ventilating wood alone is therefore unlikely to control wood-boring insects, and indeed deathwatch beetle and furniture beetle are found well away from embedded ends in well-ventilated wood.

Light traps

This method of control is based on research which shows that deathwatch beetle fly at temperatures of around 17^0C and that they are attracted to light. Churches which were investigated by PRL and BRE had temperatures of 6–12^0C, so that movement of deathwatch beetle would have been minimal and they would not have flown. There was no evidence of beetles at windows.

Experiments carried out in roofspaces during the Woodcare project tended to be warmer and good results were obtained. As a result of this the light trap is now a serious and cost-effective 'natural' approach to the control of deathwatch beetle in a suitably warm environment. But there remain many situations where this form of trap will be impractical.

Wood preservatives

In the UK all pesticides, pesticide products, labels and their use are strictly controlled by legislation under the Food and Protection Act (FEPA) and the Control of

Table 1 Equilibrium moisture contents.

External air temperature (°C)	Relative humidity (%)	Vapour pressure (kPa)	Wall temperature (°C)	Relative humidity at wall temp (°C)	Equilibrium moisture content of wood (%)
20	55	1.28	14	80	18.3
18	55	1.13	14	71	15.8
16	55	0.99	14	62	13.4

Table 2 *Relative toxicities of assorted chemicals.*

Chemical	Lethal dose (LD50) (mg/kg body weight)
Pentachlorophenol (PCP)	27
Lindane	88
Tributyltin oxide (TnTBTO)	200
Cypermethrin	251
Trichlorophenol (TCP)	820
Asprin	1000
Organo-boron ester	1700
Sodium chloride (kitchen salt)	3000
Acetic acid	3500
Boric acid	3000–4000
Sodium hydrogen carbonate ('bicarb')	4220
Permethrin	4570
Boron/glycol preservatives	8000–15000
Sugar (sucrose)	29000

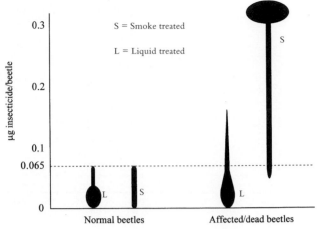

Figure 1 *Pick-up of insecticide vs conditions of beetles. See also Colour Plate 7.*

Pesticides Regulation (COPR). Under this legislation all aspects of pesticide formulation, their use and their effect on the environment is assessed by the government's independent Advisory Committee on Pesticides through the Health & Safety Executive. This provides what are probably the most strict controls and legislation on the sale and use of pesticides in Europe and possibly the world. This explains why we have so few incidents. Broad comparisons of lethal doses of preservatives and common substances are shown in Table 2. The LD50 dose is the dose at which 50% of the test animals are killed by the substance. Most products now used in wood preservatives are, at worst, labelled 'irritant' and are well down in the list (Table 2), so pesticides may not warrant the extreme caution in their use than has been the case.

How effective are chemicals against deathwatch beetle?

In the 1960s and 1970s two types of treatment were investigated by BRE and PRL in about ten sites under field conditions. Both treatments were based on contact insecticides.

Liquid sprays
Lindane in an aliphatic solvent (white spirit or odourless kerosene) was sprayed onto the surface of infested timber.

Smokes
Pyrotechnic formulations which, when ignited, released fine particles of contact insecticide into spaces, eg the body of a building. These were designed to reach sufficient heights to land on exposed timbers infested with deathwatch beetle. Because of the behaviour of the insects as then understood, good potential contact with smoke residues was expected. The smokes could be timed to correspond with the time adults potentially spent on the surface of the timber before egg laying and the period in which larvae might crawl to disperse.

Smoke treatments were however, a blanket treatment, and all exposed timbers were exposed to the chemical no matter if there was beetle infestation or not. The smoke deposits (active insecticide) were short-lived and the pattern of treatment therefore needed to be annual treatments at the onset of emergence and often two

treatments at one-month intervals. The advantage of smoke treatment was that it was inexpensive, and scaffolding was not required for access to the timbers.

Beetles which fell from the roof timbers were collected weekly, and sent to the laboratory for analysis in order to establish how insecticide residues were picked up by beetles and their subsequent effect.

The amount of insecticide contained within each beetle was measured, and the final state of the beetle assessed as normal/alive or affected/dead. Normality was a subjective judgement based on increased activity if the beetles were heated up a little. Beetles were also sexed, and the females were dissected to see whether they had mated and laid eggs.

Figure 1 shows that below a level of 0.065:g insecticide/beetle, almost all of the beetles survived in the smoke-treated areas. Above this level, beetles were affected/died (within 24 hours). Beetles from roofs which had been spray treated mostly contained residues which were below the threshold level. Smokes were much more effective than spray treatments, and the majority of beetles from areas treated with smoke picked up insecticide levels which were substantially in excess of the threshold level. Death in beetles from spray treated sites seemed to be natural mortalities while in smoke-treated sites

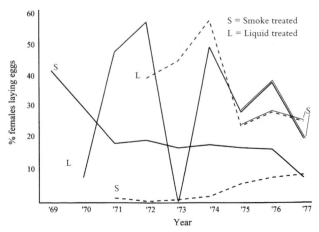

Figure 2 *Emergence of beetles 1969–77. The lower two lines show smoke-treated sites, and the upper two zigzag lines show liquid-treated sites.*

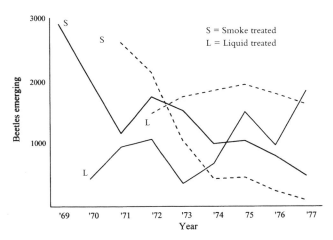

Figure 3 Egg-laying 1969–77, shown by years against 1000s. See also Colour Plate 8.

death was caused by insecticides. High levels of smoke treatments affected more beetles than did lower levels.

Figure 2 illustrates egg laying statistics for smoke- and liquid-treated sites. The numbers of beetles which had laid eggs in smoke-treated sites (ie numbers of dissected females which had laid eggs) were usually below 20% and frequently below 10% of the sample collected. On liquid-treated sites this was typically over 50%, suggesting that the insecticide deposits had had their effect before the eggs were laid. These results indicate that the smokes killed the beetles before they could lay their eggs. Many more of the beetles in liquid-treated sites had laid their eggs, therefore suggesting that death was due to normal mortality. In 1973 the drop in the percentage of females laying eggs was due to repeated spraying prior to emergence. Figure 2 illustrates the percentage of females laying eggs following a change from liquid to smoke treatment.

Figure 3 illustrates beetle emergence between 1969 and 1977. Emergence showed a clear steady decline with smoke treatment (3000–400 beetles in nine years). Liquid treatments, however, showed no long-term noticeable effect.

In late 1970s, specially formulated permethrin smokes were made for the experiment. These formed high plumes with high yields of insecticide. This meant that higher roof spaces could be accessed. Two sites were tested where permethrin smokes followed lindane treatments. Significant reduction in mating and egg laying was shown following the use of permethrin.

Emergence figures for one of the sites tested with permethrin showed a clear reduction in emergence. This is probably due to permethrin being more toxic to wood-boring insects and more persistent. The chemical also seems to stimulate activity so that beetles fell off the wood before mating and egg laying. (Beetles are frequently found on the floor even when there has been no treatment of their environment.)

Why did liquids fail?

HCH (lindane) is volatile and probably disappeared rapidly from timbers before it could be absorbed. The light surface spray gave poor initial loading and blanket treatments were not targeted at problem areas.

Smokes proved very effective but are also a blanket treatment and are therefore not appropriate for sensitive environments.

For success with liquids the wood must be treated thoroughly only where there is activity, so it is essential to identify infested areas. In cold buildings, beetles drop to the floor and remain where they fall, so areas of activity can be identified. This can be done by collecting beetles from March to June during the emergence season and marking their position on a plan. Two years of collection is normally sufficient to identify patterns of infestation and activity.

Carefully targeted, the following treatments should prove effective.

Solvent-based treatments

Solvent-based permethrin, is effective, but can lead to increased fire hazard and odour.

Boron/glycol

Sodium octaborate/glycol can be injected or applied to the surface. It will diffuse well, particularly in damp wood, but is also quite good at diffusing in drier woods, especially where the relative humidity is high. The chemical acts as a stomach poison or antifeedant and renders wood unsuitable for larval activity.

The benefits are that the treatment has a low environmental impact, is odourless and has superb fungicidal activity so that it is good for potentially damp, embedded timbers.

'Paste' preservatives

Inverted oil-based emulsions containing permethrin have a high oil content, which means that the paste penetrates well into 'dry' wood. A disadvantage is that the oily residues may stain adjacent plaster and decoration. Paste preservatives can be applied by injection as for boron/glycol or applied as surface treatment.

CONCLUSIONS

Chemical treatments in the form of smokes have proved effective in reducing numbers of deathwatch beetles. Unfortunately suitable formulations which produce a plume height sufficient to reach high timbers are not currently available.

Liquids have failed to control deathwatch beetle in the past due to their inability to concentrate and target infected areas. The key to success with liquids is planned targeted treatments.

Chemical treatments should be considered as part of an integrated control programme with other methods.

Finally, even if total eradication of an infestation is not achieved, a significant reduction in beetle activity is likely to prolong the life of historic timbers.

AUTHOR BIOGRAPHY

Graham Coleman has qualifications in zoology, chemistry and botany, and was formerly a Senior Scientific Officer at the Building Research Establishment. He now works in private practice specialising in laboratory and consultancy services examining dampness and timber infestation in buildings.

Part II

Fungi

Recent studies on the oak polypore *Donkioporia expansa* (Desm.) Kotl. & Pouz.

Colm P Moore and Hubert T Fuller[*]

Department of Botany, University College Dublin, Belfield, Dublin 4, Ireland;
Tel: + 353 1 716 2341; Fax: +353 1 716 1153; email: hubert.fuller@ucd.ie

Abstract

Donkioporia expansa is a destructive white-rot fungus found on oak timbers in buildings and is sometimes associated with deathwatch beetles. Cultural and growth characteristics of twelve European *Donkioporia* isolates are compared and the relationship of isolates discussed. The effects of temperature, pH and water availability on mycelial growth are described. For most isolates growth was optimal at 25°C to 30°C and at pH 4.1 to 4.4. Growth was inhibited at water potentials in the range −6.0 to −9.0 MPa. Chlamydospore germination and survivability was also studied and the role of chlamydospores as dispersive propagules is discussed. Degradative enzymes secreted by *Donkioporia* were identified. In a series of wood decay tests, the maximum weight losses observed due to *Donkioporia* growth on oak heartwood blocks were in the range 5–10 %. Results indicated that prolonged exposure to high moisture conditions is required for substantial decay of oak heartwood. Boron-based fungicides inhibited *Donkioporia* growth.

Key words

Donkioporia, white-rot fungus, oak

INTRODUCTION

Donkioporia expansa is a destructive, wood-rot fungus usually found on decaying oak timbers in old historic buildings in Europe. Apart from its notoriety as an oak-rotting fungus *Donkioporia* is also known because of its occurrence with deathwatch beetles. The simultaneous presence of fungus and beetle in decaying oak wood has led to investigations of a possible relationship between fungus and beetle (EU Woodcare Project 1994–97). The present studies on *Donkioporia* were undertaken to provide information on the growth and physiology of the fungus, in support of deathwatch beetle investigations (Donnelly 1997, Belmain et al this volume, Ridout & Ridout this volume).

The extensive destruction of structural timbers in the Palace of Versailles in the early part of the twentieth century established *Donkioporia*, then known as *Phellinus cryptarum*, as an important wood-rotting fungus (Mangin & Patouillard 1922). Since then there have been numerous other cases of *Donkioporia* rot in oak-structured historic buildings such as cathedrals, castles, municipal buildings and residences. Cartwright and Findlay (1936, 1958) claimed that *Donkioporia* (*Phellinus megaloporus*), more than any other fungus, had the potential to cause rapid, significant decay of oak. More recently, Buchwald stated that 'this fungus has a considerable importance as a wood-destroyer in buildings' (1986, 2). Once considered rare, perhaps because it was not recognised, there is now greater awareness of the occurrence of *Donkioporia* in Europe: in Czechoslovakia (Kotalba & Pouzar 1973, Kotalba 1984), Germany (Jahn 1967, Dörtfelt & Sommer 1973, Buchwald 1986, Ritter 1992, Kleist & Seehann 1999), Belgium (Guillitte 1992, Decock & Hennebert 1994), The Netherlands (Esser & Tas this volume) and Great Britain (Ridout pers comm, White *et al* 1996, Ridout 2000, Ridout & Ridout this volume). It has also been found in Italy, France, Hungary and Switzerland (Heim 1942, Wazny & Czajnik 1963, Ryvarden & Gilbertson 1993, Szabo & Varga 1995). After *Serpula lacrymans*, *Donkioporia* was one of the most common fungi observed on samples of decayed wood in Belgium (Guillitte 1992, Decock & Hennebert 1994). Similarly, in Germany, Kleist and Seehann (1999) noted that while dry rot accounted for wood decay in 40% of samples examined at the Institut für Holzbiologie in Hamburg since 1995, 'Eichenporling' (*Donkioporia*) was responsible for 20% of wood rots. So far, the fungus has not been found in Scandinavian countries or Ireland, giving the fungus a sub-Atlantic/Central European distribution. There has been one report of the fungus (as *Poria megalopora*) in Siberia (Bondartsev 1971) and sporadic finds in Canada (Ontario) and the USA (Ohio & Illinois) (Baxter 1950, Gilbertson & Ryvarden 1986, Gessner pers comm).

Donkioporia is found on worked oak timbers but it also causes decay of other hardwoods such as chestnut, cherry, ash, poplar, hickory, and of conifer woods such as spruce, larch and pine (Ritter 1988, Ryvarden & Gilbertson 1993, Szabo & Varga 1995, Kleist & Seehann 1999). According to Buchwald (1986) oak wood was the most frequent substratum, representing 42% of wood samples, while coniferous timbers (spruce and pine) accounted for 54%. The fungus probably has a wider substratum range as it can colonise other wood types *in vitro* (Buchwald 1986) and also durable tropical woods in particular circumstances (Van Acker & Stevens 1996). It has most frequently been found in buildings, growing in roofs, cellars and outhouses, on trusses, tie beams, rafters,

Figure 1 Donkioporia *white rot of oak wood, showing fibrous nature of decayed wood. Bar = 20 mm.*

flooring and panelling. Less frequently it has been found externally on window and door frames, wooden bridges and fence posts (Jahn 1967, MUCL 1998). More unusual locations include a nineteenth-century wooden frigate (White *et al* 1996) and cooling tower timbers (Van Acker & Stevens 1996). Like *Serpula lacrymans*, the original natural habitat of *Donkioporia* is a matter for debate. Ryvarden and Gilbertson (1993) noted that *Donkioporia* occurs on fallen trunks of *Quercus* in warm and dry forests. No source was cited by Ryvarden and Gilbertson but Kleist and Seehann (1999) suggest that the comment may refer to the polypore records of Baxter (1950) from the region of the Great Lakes in North America. One of the herbarium specimens examined by Kotalba and Pouzar (1973) was from a fallen trunk of *Quercus cerris* in central Slovakia and in Hungary the fungus is known locally as 'chestnut tinder', perhaps indicating a natural niche in that region. Ellis and Ellis (1990), also cited in Watkinson (1994), listed *Donkioporia* as being uncommon and occurring mostly on wood of *Castanea* and *Quercus*. In the context of their publication this implies that it grows on these substrata in nature, but no supporting evidence was given. Interestingly, one of the *Donkioporia* isolates in the MUCL culture collection was isolated from a fence post close to *Quercus* woodland in France (MUCL 1998).

Since first being described and named as *Boletus expansus*, by Desmazieres in 1823, the oak rot fungus has undergone numerous nomenclatural changes. The existence of twelve synonyms for the fungus reflects the difficulties experienced by mycologists in finding a satisfactory generic position for the taxon. The fungus has been variously known as *Poria megalopora* Pers. Cooke 1886, *Phellinus cryptarum* Auct. dt. Karst, *Phellinus megaloporus* (Pers.) Heim 1942 and *Poria expansa* (Desmaz.) Jahn 1976, among others. In 1973, Kotalba and Pouzar proposed a new monotypic genus, *Donkioporia*, to accommodate this problematic species, which they named in honour of Dutch mycologist, Dr Marinus Donk. *Donkioporia expansa* (Desmazieres) Katalba et Pouzar 1973 is now generally accepted as the name for the fungus. It is a Homobasidiomycete, classified to the family Polyporaceae (Ryvarden & Gilbertson 1993).

DONKIOPORIA IN SITU

Donkioporia is a wet-rot fungus which grows best and develops where there is persistent moisture, often in wood

Figure 2 Donkioporia *resupinate basidiocarp on surface of oak beam showing advanced white-rot in centre. Bar = 40 mm.*

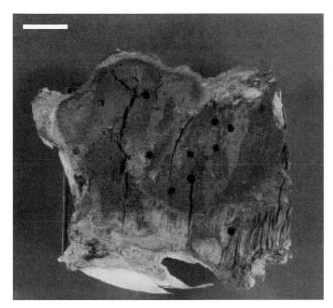

Figure 3 Fruiting body of Donkioporia *showing boreholes of* Xestobium rufovillosum. *Bar = 12 mm.*

subjected to wetness over a prolonged period. Beam-ends within damp walls are particularly prone to attack. The rot produced by the fungus is a white, fibrous stringy rot (Fig 1), which seriously reduces the mechanical strength of wood. Decay may be well advanced within timber beams before the fungus becomes apparent as differentiated basidiocarps/ sporophores on the wood surface (Fig 2).

In some cases, oak timber colonised by *Donkioporia* is also infested by larvae of deathwatch beetle (*Xestobium rufovillosum*) (Mangin & Patouillard 1922, Fisher 1940, Ridout pers comm). Boreholes are also occasionally found in *Donkioporia* basidiocarps (Fig 3). Fisher stated that 'the presence of this fungus in a building has an important bearing on the occurrence and severity of *Xestobium* damage'. Based on chemical analyses of de-

cayed oak wood and also on feeding experiments, Campbell and Bryant (1940) concluded that the decline in mechanical strength of decayed wood, caused by any fungus, is the major factor facilitating penetration by beetles and their larvae. They considered the possibility that fungal (*Donkioporia*) mycelium might contain 'a component essential to the beetle', but none was detected in their analyses. A relationship between fungus and beetle has long been suspected and the testing of this hypothesis was a primary objective of the Woodcare Project (Belmain et al this volume, Ridout & Ridout this volume).

Basidiocarps of *D. expansa* are flat, pulvinate, resupinate and poroid. They take the form of broad, spreading plates, 100–300 mm (4 – 11 in) wide (Figs 4 and 5), but are known to be capable of covering several square metres, as the fungus grows over and around timber surfaces (Fig 6). Basidiocarps may take on a bracket form where they overgrow surfaces. Strand formation is not known to be a feature of *Donkioporia* on wood in buildings (Cartwright & Findlay 1958, Kleist & Seehann 1999) but strand-like structures have been observed by the authors under some conditions *in vitro*. In thickness, basidiocarps taper from *c* 30 mm (*c* 1 in) at their centre to a few mm at the growing margin. They are perennial and several successive tube layers may be present. Colours vary through shades of light brown to grey brown to fawn/cinnamon, with a velvety appearance. Wet sporophores tend to be darker brown in colour. Fresh, actively growing sporophores are tough, spongy and elastic while older, drier specimens are woody, light and brittle. Basidiocarps are relatively easily detached from their substrata. A black melanized layer may be seen beneath basidiocarps where they attach to the substratum. Macroscopically, *Donkioporia* basidiocarps are similar to other poroid resupinate

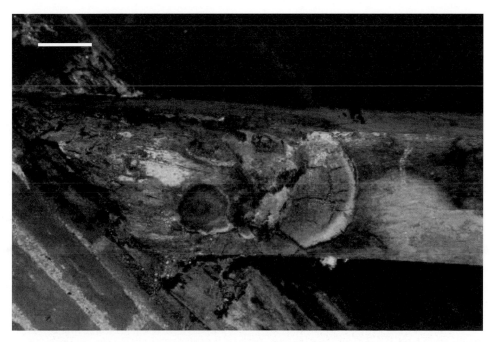

Figure 4 Donkioporia *basidiocarp (fruiting body) on oak roof timbers at Stoneleigh Abbey, Warwickshire, UK (April 1997). Bar = 50 mm.*

Figure 5 Donkioporia *basidiocarp on oak panelling at Stoneleigh Abbey, Warwickshire, UK (April 1997). Bar = 25 mm.*

Figure 6 Spreading basidiocarp of Donkioporia *on oak beam in roof at Stoneleigh Abbey, Warwickshire, UK (April 1997). Bar = 100 mm. See also Colour Plate 9.*

fungi (eg *Phellinus contiguus, Perenniporia medulla-panis* and some spp of *Gloeophyllum*). Detailed distinguishing macro- and microscopic features of *Donkioporia* are given in Jahn (1967, 1971), Domanski and Orlicz (1967), Domanski (1972), Kotalba and Pouzar (1973), Ritter (1983), Julich (1984), Ryvarden and Gilbertson (1993), Kleist & Seehann (1999).

DONKIOPORIA IN CULTURE

Isolates of *D. expansa* were either acquired from culture collections or isolated from basidiocarps obtained from degraded constructional oak timbers in buildings (Table 1). The basidiocarps were supplied to the authors by Ridout Associates (Stourbridge, England) and pending deposition in European culture collections the isolates have been temporarily designated 'Rid. Coventry', 'Rid. Devon' and 'Rid. Stoneleigh'.

Stock cultures of all isolates were maintained at 4°C on potato dextrose agar slopes and also on sealed deep plates. A further collection was maintained on agar plugs in sterile water. Stock subculturing was performed every six to nine months. As most isolates were in regular usage, isolate purity and stability was constantly monitored.

Table 1 Donkioporia *isolates used in current research project.*

Donkioporia expansa isolate designation	source
NCWRF 178E [1]	Unknown, United Kingdom, 1952
NCWRF 178F	Sporophore in church, United Kingdom, 1954
CBS 236.91 [2]	Wood, *Quercus* sp., Germany, 1991
CBS 299.93	Antique oak chest, Netherlands, 1993
CBS 509.89	Forest soil, Germany, 1989
MUCL 30747 [3]	Bore dust of *Xestobium rufovillosum*, *Quercus* sp., Plombiere, Belgium, 1990
MUCL 30819	Timber, *Quercus* sp., house, Ham-sur-Heure, Belgium, 1990
MUCL 31565	Bore dust of *Xestobium rufovillosum*, *Quercus* sp., house, Brussels, Belgium, 1991
MUCL 31593	Floor, conifer, house, Saint-Gilles, Brussels, Belgium, 1991
Rid. Coventry	Sporophore on timber, *Quercus* sp., Courthouse, Coventry, England, 1996
Rid. Devon	Sporophore on timber door frame, *Quercus* sp., Poltimore House, Exeter, Devon, England, 1997
Rid. Stoneleigh	Sporophore on roof beam, *Quercus* sp., Stoneleigh Abbey, Warwickshire, England, 1997
Var 110a	Sporophore on old roof beam timbers, Norway spruce, Sopron, Hungary, 1991 (donated by Dr Ferenc Varga)

[1] National Collection of Wood-rotting Macro-fungi, Building Research Establishment Department of the Environment, Watford, UK
[2] Centraal Bureau voor Schimmelcultures, Baarn, The Netherlands
[3] Mycotheque de L'Universitie Catholique de Louvain-la-Neuve, Louvain, Belgium
MUCL maintains a sizable collection of *Donkioporia expansa* isolated by Dr C Decock, including some from *Xestobium*-infested oak timbers.

Particular care was necessary with Isolate 299.93 because of its tendency to sector; choice of a 'rogue' sector could lead to loss of the original type. Older plate cultures of *Donkioporia* were prone to invasion by mycophagous mites and particular vigilance and care was required to prevent such attacks. Stringent anti-mite precautions were necessary, miticides periodically used, and cultures routinely sealed. The attractiveness of the cultures to mites may indicate the emission of compounds detectable by feeding insects.

The availability of a collection of *D. expansa* isolates afforded the opportunity to compare their colony morphologies and other characteristics (Moore 2001). At present there is no definitive description in the mycological literature of *D. expansa* in culture. Several authors have described individual or a few *Donkioporia* isolates (Cartwright & Findlay 1958, Domanski & Orlicz 1967, Ritter 1983) and based on their descriptions and the variation observed between isolates examined in this

study, a narrow exclusive species description does not seem feasible.

One-week-old colonies on potato dextrose, malt extract, cherry agar (Stalpers 1979) or comparable media, at 25°C, consist of white/off-white, spreading, low-growing mycelia (Fig 7). Growth rates are relatively fast (c 4–5 mm day^{-1}). Production of aerial hyphae varies between isolates. In most, a downy felt of short hyphae is formed, whereas in others (eg Isolates 178E & 30819) hyphae grow close to the surface or are mostly submerged, resulting in smooth-surfaced colonies. A sweet, fruity, aromatic smell is a feature of most actively growing cultures. In contrast, Domanski and Orlicz (1967) noted that Isolates 17E and 178E both had an 'unpleasant smell of dishwater'!

With further growth of the colonies, textural and other differences between the isolates become apparent (Fig 8). In a few isolates aerial mycelial development is patchy and thin, giving a transparent look to the colonies (eg Isolate 30819). In others (eg Isolate 30747) the texture of the mycelial mat is variously floccose, consisting of hyphal tufts. These may be small and localized giving a grainy appearance, or they may be more widely distributed and conjoined over much of the colony surface. A few isolates are plumose, with radiating raised spoke-like fingers of hyphae (Isolate 236.91). Isolate 299.93 is more densely floccose/woolly with long radiating prostrate hyphal bundles. Concentric zonations, from few to many, which may indicate a sensitivity to light, are also evident in several isolates.

With age, the white mycelia of all isolates deepen to off-white/cream/buff colorations. Isolate 31565 is distinctive in that, with time, the colony darkens brown from the centre out. This darkening is also seen in the reverse reddish-brown coloration of the colony. With age the reverse colorations of all isolates darken to varying degrees. The light yellow-orange coloration of Isolate 299.93 and the intense yellow reverse coloration readily distinguish this isolate from others. Red-brown to dark brown exudates are formed on the mycelial mats of several isolates, especially 30747 and 31565. Initially small droplets appear, which are associated with lacunae

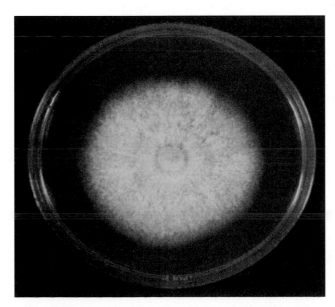

Figure 7 Five-day-old colony of Donkioporia expansa *(Isolate 31565) growing on potato dextrose agar. 90 mm Petri dish.*

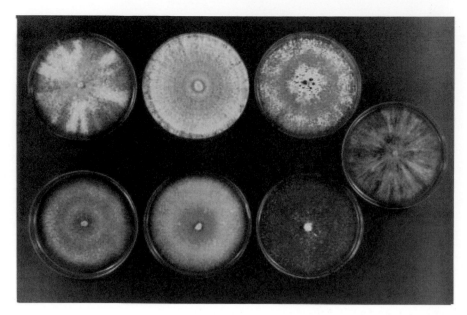

Figure 8 Variation in colony morphologies of two-week-old Donkioporia *isolates on potato dextrose agar. From top left, clockwise: 236.91, 299.93, 30747, 31565, 30819, 178E and 31593. 90 mm Petri dish. See also Colour Plate 10.*

in the mycelium and these droplets eventually merge to form large blobs, usually at the centre of plates (Fig 9). Similar droplets have been observed on basidiocarps (Ritter 1988, Kleist & Seehann 1999) and also on mycelium on wood, by the authors.

After one to two weeks, differentiation becomes evident in the mycelial mats of most isolates. Erect dissepiments develop and these join, leading to the formation of a flat resupinate poroid surface, usually around the inoculum plug or at the periphery of dishes. The poroid tissue may eventually spread over most of the colony surface (Fig 10) and a mushroom-like smell can be detected from these cultures. Differentiation is readily induced, on a variety of natural and defined media, liquid or solid, in light and dark and at different temperatures. Mycelial extracts of *Donkioporia* were found to induce

earlier and more extensive differentiation (O'Shea 1996), a 'self' phenomenon also noted in *Phellinus* by Butler (1995). Basidiome formation in *Donkioporia* cultures has also been observed by Cartwright and Findlay (1958), Domanski and Orlicz (1967) and Ritter (1983). Hegarty and Buchwald (1988) produced *Donkioporia* fructifications using a technique developed for *Serpula lacrymans* (Cymorek & Hegarty 1986).

The differentiated poroid tissue varied in colour from deep cream to buff to cinnamon. The extent of differentiation also varied from isolate to isolate, prolific in 30747 to sparse in 299.93. The tissue was constructed of erect dissepiments, 1.5–2.0 mm high, which fused to form pores of different shapes, initially labyrinthine but later as more dissepiments merged, the shapes became angular to round (Fig 11). Pore diameters and densities varied

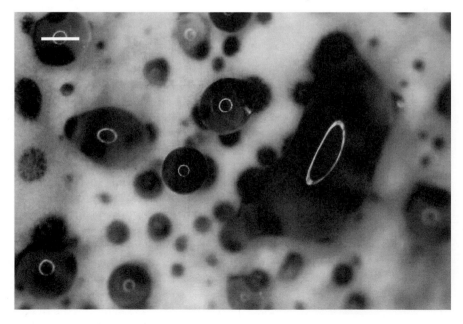

Figure 9 Exudate production by Donkioporia *(Isolate 30747) growing on potato dextrose agar. Bar = 2 mm.*

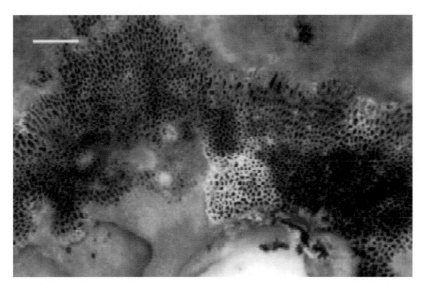

Figure 10 Culture of Donkioporia *(Isolate 30747) showing poroid formation at the edge of a Petri dish. Bar = 10 mm. See also Colour Plate 11.*

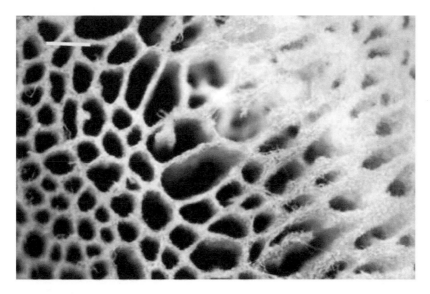

Figure 11 In vitro *differentiation of dissepiments varying from angular to round in shape, of* Donkioporia *on potato dextrose agar at 25°C. Bar = 1 mm. See also Colour Plate 12.*

between isolates, ranging from 15–40 μm and 7–11 pores per mm², respectively. The poroid context formed *in vitro*, either on culture media or on wood blocks, bore little resemblance in appearance and texture to naturally occurring basidiocarps in buildings. For example, pore diameters and densities in basidiocarps *in vivo* are 90–250 μm and 4–5 per mm, respectively (Domanski 1972). In the present study attempts at stimulating basidiocarp formation on large oak blocks incubated under wet conditions resulted only in the formation of aberrant poroid tissue. Rayner and Boddy (1988) state that fruiting structures formed by some basidiomycetes in culture, under certain circumstances, are morphologically atypical. Accordingly, reliance on features of *Donkioporia* basidiomes formed *in vitro* for identification purposes could be misleading.

Basidiospore production by *Donkioporia in vitro* appears to be erratic. Cartwright and Findlay (1958) failed to find any, while Domanski and Orlicz (1967), Ritter (1983) and Hegarty and Buchwald (1988) have observed basidia and basidiospores on poroid structures *in vitro*. In the present study small deposits of basidiospores (Fig 12) were noted on the lids of inverted dishes containing differentiating colonies of Isolate 30747.

Hyphal features, chlamydospore formation and the mitic systems of poroid structures were also examined. Clamp connections were observed on the undifferentiated hyphae of all isolates, though infrequently in some. Where present they could often be found at branch junctions. The poroid tissue was trimitic, with generative, skeletal and binding hyphae present. Of the twelve *Donkioporia* isolates examined, all but one (Isolate 299.93) readily produced chlamydospores when grown on a variety of culture media (Fig 13) and on wood. An estimated 10³–10⁴ spores were formed per mm² of culture surface area. Chlamydospores have also been observed present in

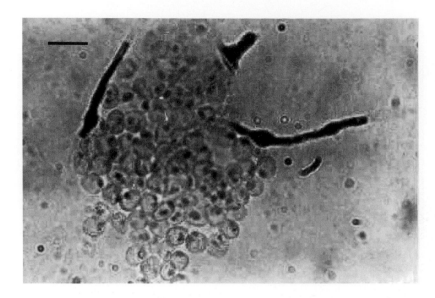

Figure 12 Basidiospores of Isolate 30747 stained with lactophenol acid fuschin. Bar = 10 μm.

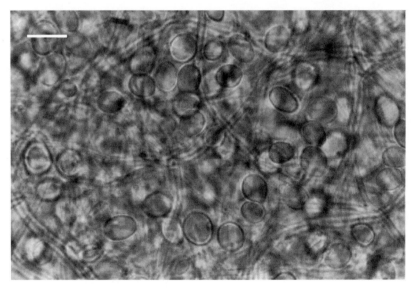

Figure 13 Chlamydospores in situ *on the surface of a two-week-old culture of* Donkioporia expansa *(Isolate 31565) on potato dextrose agar. Bar = 10 μm.*

differentiating basidiomes *in vitro,* in basidiocarps taken from timbers in buildings and in colonised wood. Most authors, from Cartwright and Findlay (1958) to Szabo and Varga (1995) have noted the presence of chlamydospores in their descriptions of *Donkioporia*. A few other wood-inhabiting fungi of the Polyporaceae group are also known to produce chlamydospores (Stalpers 1979, Breitenbach & Kranzlin 1986, Ryvarden & Gilbertson 1993).

Chlamydospores are thick-walled intercalary thallic conidia. They are formed within vegetative hyphae by local accumulation of protoplasm and septum formation and are released for dispersal when the hyphae desiccate and fragment. Dispersal may be by air movements or facilitated by mycophagous and other wood-inhabiting insects. Chlamydospores are generally considered as resting spores, enabling dispersal and survival when environmental conditions are unfavourable for mycelial growth. On the other hand, basidiospores produced in basidiocarps

are short-lived, thin–walled spores, prone to desiccation and unlikely to function as survival structures.

In *Donkioporia*, chlamydospores are produced either singly or in short chains. At room temperature they form within days and appear like beads along the hyphae (Fig 14). They are generally ovoidal in shape, measuring *c* 10 ± 1.2 x 7.5 ± 0.6 μm, with a smooth, thick dark wall (*c* 0.25 μm) without ornamentation and a granular cytoplasm, hyaline to straw-coloured. They readily disarticulate from the mycelium and may be seen to carry residual wall fragments.

Nobles (1965), Stalpers (1979) and later Nakasone (1990) used numerical species codes in the identification of basidiomycetes in culture. This approach was investigated using the morphological and other characters collated for the various *Donkioporia* isolates. Of the twelve isolates described, only the following had identical codes using the Nakasone system: MUCL 31593, Varga 110a,

Figure 14 Scanning electron micrograph of Donkioporia expansa *showing chains of chlamydospores in situ. Bar = 10 μm.*

Rid. Coventry, Rid. Devon and Rid. Stoneleigh (Moore 2001). The code determined for these isolates was: 2, 3, 8, 13, 16, 24, 34, 37, 39, 43, 48, 53, 54/55. Other isolates differed in one or few cultural features and accordingly had different codes. Domanski and Orlicz (1967) and Ritter (1983) also determined species codes for individual *Donkioporia* isolates. Domanski and Orlicz's codes do not totally correspond with any of Moore's, while Ritter used the Stalpers (1979) approach, which differs from the other coding systems.

The foregoing results point to the need for further critical work on the taxonomy of *Donkioporia expansa*, both as basidiocarps found on wood and as cultures *in vitro*. *Donkioporia* isolates from Belgian, Dutch and British culture collections and those cultured from basidiocarps during the project, exhibited considerable variation in colony morphology when grown *in vitro* (Fig 8). While some isolates had features in common, others were distinctly different, suggesting the existence of different ecotypes or even species. Morphological and growth criteria of cultures were unsatisfactory in clarifying the relatedness of the isolates. Compatibility tests using monokaryon crosses were not possible because basidiospores were unavailable from which to establish haploid cultures. Dual culture interactions were studied, which involved confronting each isolate with self and also with each of the other isolates. Some authors (Sharland & Rayner 1986) have used this method to determine incompatibilities and the closeness of relationships between fungal isolates. Various interaction patterns (intermingling, translucent, 'wam' *etc.*) were observed but no conclusions could be reached in relation to the genetic closeness of the *Donkioporia* isolates.

The genetic relatedness of *Donkioporia* isolates was further investigated using several DNA-based methods. Mitochondrial DNA polymorphism was examined but considerable difficulty was experienced in isolating adequate amounts of intact mitochondrial DNA. Another approach which failed to give satisfactory results involved developing a multilocus probe for hybridization and DNA fingerprinting (Peters 1996). DNA polymorphisms were also investigated using RAPDs. Using a collection of primers, preliminary RAPD results were promising, indicating four distinct sub-populations among the ten *Donkioporia* isolates examined (Green 1998). Relationships will be further clarified as more primers are used and additional isolates are included. Recently Moreth and Schmidt (2000) designed primers which were specific for variations in the ITS region of ribosomal DNA from a range of wood-rotting fungi. The PCR profiles obtained using these primers were successfully used in the identification of *Donkioporia* and other fungal species in rot samples from buildings.

IN VITRO GROWTH OF *DONKIOPORIA*

Radial growth and biomass production

All isolates of *D. expansa* readily grew and produced dense mycelial colonies on a variety of commonly-used mycological agar media eg malt extract (MEA), potato dextrose (PDA), cornmeal agar. Cherry agar, used by Stalpers (1979) for comparing growth of basidiomycetes in culture, also supported good growth, as did the defined Raulin's (modified) medium used by Bassett *et al* (1967) for *Heterobasidion*. Growth on a defined medium lacking thiamine yielded sparse growth, suggesting a requirement for this vitamin. While growth was obtained using inorganic N sources, organic sources (eg asparagine, peptone) enhanced mycelial growth. Media supplemented with wood shavings or sawdust supported excellent growth and mycelial growth was also possible on cellophane-overlaid media. Both PDA and MEA were used successfully when attempting to establish *Donkioporia* cultures from basidiocarp tissue. Antibiotics were not necessary in the isolation medium,

Table 2 *Radial growth rates (mm day⁻¹) of twelve* Donkioporia *isolates, of varying geographical origins, on natural (PDA & MEA) and synthetic (Raulin's, modified) media at 25°C. Values represent means of three replicate growth rates.*

Isolate	Geographical origin	Potato dextrose agar	Malt extract agar	Raulin's agar
NCWRF 178E	United Kingdom	4.3	4.0	3.8
NCWRF 178F	United Kingdom	3.4	4.1	–
CBS 236.91	Germany	4.1	4.5	–
CBS 299.93	Netherlands	6.5	5.5	3.5
MUCL 30747	Belgium	4.2	4.6	4.4
MUCL 30819	Belgium	3.9	3.7	4.3
MUCL 31565	Belgium	4.1	3.4	4.3
MUCL 31593	Belgium	4.1	4.6	4.5
Rid. Coventry	United Kingdom	–	4.9	–
Rid. Devon	United Kingdom	–	4.2	–
Rid. Stoneleigh	United Kingdom	–	4.6	–
Var 110a	Hungary	3.8	4.8	–

as fungi (eg *Trichoderma*), rather than bacteria, were the usual contaminants.

Table 2 lists the radial growth rates (K_r mm day⁻¹) at 25°C for *Donkioporia* isolates on PDA, MEA and Raulin's (modified) agar. It is not uncommon to find growth rate differences on different media, especially between a natural and synthetic medium. Radial growth rates on PDA ranged from 3.9 mm day⁻¹ for Belgian Isolate 30819 to 6.5 mm day⁻¹ for Dutch Isolate 299.93. Such differences in growth rates are commonly found for fungal isolates of the same species from different geographic regions.

Ritter (1983) recorded a growth rate of 3.5 mm day⁻¹, at 20°C, for a German (Madgeburg) isolate. Cartwright and Findlay (1958) gave a value of 8.5 mm day⁻¹, at 27°C, which may be a diameter rate. Working with a Hungarian isolate of *Donkioporia* Szabo and Varga (1995) reported a growth rate of 4.4 mm day⁻¹ on MEA at 25°C, which corresponds favourably with the West European isolates. Growth rates are such that at 25°C a standard 90 mm Petri dish is filled by *Donkioporia* within nine to twelve days, which is comparable to other wood-rotting fungi. Rate of mycelial spread *in vitro* does not necessarily reflect rates of hyphal penetration in wood or of basidiome development, which would have to be determined independently.

In order to assess biomass production, *Donkioporia* isolates were grown in liquid culture in a variety of media and culture vessels. In the first experiment isolates were grown in potato dextrose broth (PDB) in Petri dishes at 27°C. Despite variation in the biomass yields for the different isolates (Fig 15), most of the isolates grew well, some profusely, in this medium. Fungal biomass values ranged from 97–435 mg dry weight mycelium after three weeks.

In a further experiment, Isolate 178E was grown in various natural media in 250 ml conical flasks for five weeks at 25°C (Fig 16). The graphs showed sigmoidal-type kinetics typical of mycelial fungi in stationary batch culture. Stationary (plateau) phases were reached after fifteen days suggesting nutrient depletion or byproduct accumulation and staling. After twenty-one days of incubation at 25°C, biomass production was greatest in the peptone-supplemented broths, with 280 and 317 mg dry weight mycelium in the potato dextrose and malt extract supplemented media, respectively. Biomass pro-

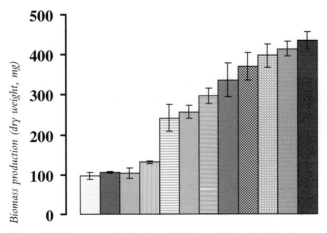

Figure 15 *Biomass production (mycelial dry weight, mg) of twelve* Donkioporia expansa *isolates in potato dextrose broth in Petri dishes at 27°C after 21 days. Bars represent standard errors of means of five replicate measurements. Isolates (L-R):178E, 30819, 178F, 299.93, 236.91, 31593, Rid. Coventry, 30747, 31565, Rid, Devon, Rid. Stoneleigh and Var 110A.*

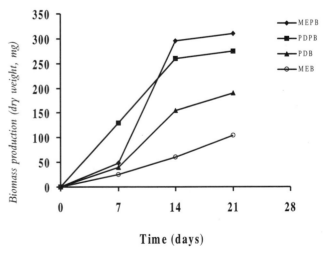

Figure 16 *Biomass production by* Donkioporia expansa *(Isolate 178E) grown at 25°C on potato dextrose broth (PDB) and malt extract broth (MEB) with and without 0.5% mycological peptone (PDPB and MEPB, respectively), in 250 ml conical flasks. Data points represent means on five replicate harvests.*

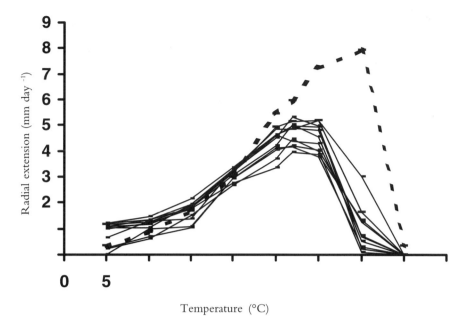

Figure 17 Influence of temperature (°C) on radial growth rates of twelve Donkioporia expansa *isolates on malt extract agar. Growth rates are slopes of regression lines for each isolate at each temperature, based on three replicate measurements. All isolates show similar trends in growth except 299.93 (dashed line).*

duction was less in the unsupplemented broths, indicating that the additional nitrogen provided by the peptone established a more favourably balanced growth medium. Fungal mycelial biomass (dry weight) was 195 mg and 107 mg, in the potato dextrose and malt broths, respectively.

In a subsequent experiment, biomass production by four other *Donkioporia* isolates (31565, 31593, 30747 & 30819) was assessed in PDP broth. Despite variation in the biomass yields for the different isolates, and in the patterns of growth, all of the isolates grew profusely in

this medium. Fungal biomass values ranged from 312 to 391 mg dry weight mycelium after five weeks, thus confirming the suitability of potato dextrose peptone as a growth medium for *Donkioporia* isolates.

In an attempt to further increase biomass yields, culture vessel volume was increased from 250 ml conicals to 1.8 L Fernbach flasks and 4.2 L Penicillin flasks. *Donkioporia* isolate 31565 was grown for five weeks in PDP broth in Penicillin flasks and in Raulin's (modified) medium in Fernbach flasks. Biomass production again followed sig-

(a)   **(b)**

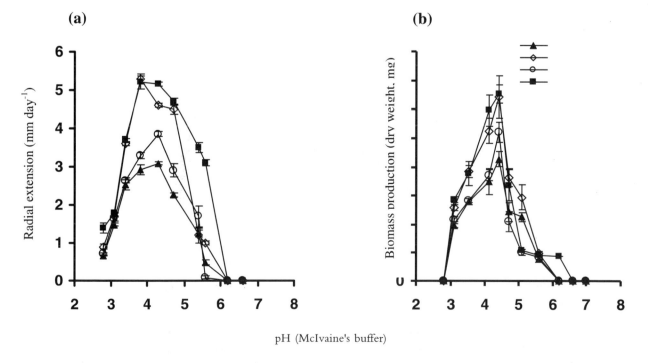

Figure 18 Effect of pH on mycelial growth as assessed by (a) radial extension (Kr, mm day⁻¹) on potato dextrose agar, and (b) biomass production (mycelial dry weight, mg) in potato dextrose broth after 21 days at 27°C. Data points (with standard error bars) represent the means of three and five replicate measurements for (a) and (b) respectively. Media were pH adjusted using McIlvaine's buffer system.

moidal-type kinetics with a stationary phase establishing by four weeks. Mycelial yields were considerably increased in the larger culture vessels. A mean dry weight of 7.4 g mycelium was harvested after five weeks in PDP broth and 5.8 g in the defined medium in Fernbach flasks; these dry weights correspond to *c* 700 g and 500 g fresh weight mycelium per flask, respectively.

The foregoing experiments provided the necessary information to bulk culture *Donkioporia* for extraction of chemical compounds (Donnelly 1997). The results also laid the basis for physiological and other studies on *Donkioporia*. The choice of culture medium, natural or defined, and the choice of culture vessel would be dictated by the requirements of the experiment. Greater mycelial yields were obtained in supplemented natural media but defined media would be preferred in some investigations.

Effects of temperature, pH and water availability

The effect of temperature on radial growth of *Donkioporia* is illustrated in Figure 17. With the exception of Isolate 299.93, the pattern of response to temperature was uniform for *Donkioporia* isolates. On the basis of these observations, *D. expansa* can be classified as a mesophilic fungus. Growth occurs between 5° and 40°C, with temperature optima *c* 25°–30°C. Cartwright and Findlay (1958) and Szabo and Varga (1995) recorded temperature optima of 27° and 28°C, respectively.

As with other fungi, it can be difficult to reconcile the laboratory data with field observations. According to Bourdot and Galzin (1928) *Donkioporia* grows only during the warm season. Jahn (1967) reported finding actively growing, fertile basidiocarps in central Germany between December and March while Ritter (1988) noted that *Donkioporia* was more commonly found in the warmer regions of central Germany (Madgeburg, Halle) than in the northern parts of the country. Szabo and Varga (1995) found a basidiocarp growing in an unheated attic in January, in Hungary. In April 1997 one of the authors (HTF) observed actively growing *Donkioporia* basidiocarps in an open attic space at Stoneleigh Abbey, England. In relation to the foregoing, sometimes conflicting observations, it is likely that the temperature and other environmental requirements for mycelial growth and wood decay are different to those necessary for basidiocarp differentiation and development. Temperatures triggering fruit body differentiation in basidiomycetes are typically lower than those favouring mycelial growth.

The effect of a range of pHs on the radial extension (Fig 18a) and biomass production (Fig 18b) of *Donkioporia* was studied using pH buffered culture media. Growth as assessed by both methods was optimal between pH 4.1 and 4.4. The results clearly show that *Donkioporia* is an acidophilic fungus and cannot establish growth even under mildly alkaline conditions. In common with other wood-rotting fungi (Cartwright & Findlay 1958), *Donkioporia* acidified unbuffered media on which it was growing. On PDA, a drop from pH 5.1 to pH 3.5 was recorded over twenty-one days. The pH of oak wood has been recorded as ranging between 3.5 and 4.5.

One of the main contributing factors favouring wood decay by fungi is high moisture and humidity. It is well established that fungal growth and wood decay are retarded in the absence of water, yet for most wood decay fungi the precise limits are not known nor are their optimum water requirements. Such knowledge is obviously relevant to the development of strategies for control of wood-rotting fungi. *Donkioporia expansa* in buildings is known to thrive in situations of high water availability. White *et al* (1996) isolated *Donkioporia* from timbers of a frigate, with wood moisture contents in the range of 112–140% while Van Acker and Stevens (1996) reported the fungus actively decaying wood with a moisture content in excess of 150%.

In a series of experiments the water relations of *D. expansa* in relation to mycelial growth were investigated (Moore & Fuller 1997, Moore 2001). Isolates were assessed in culture media of different osmotic potentials (Ψ_π) in the range -0.4 to –9.0 megapascals (MPa). The various Ψ_π were obtained by the addition of known molal concentrations of solutes such as polyols and inorganic salts (Robinson & Stokes 1955, Harris 1981) to a basal medium of PDA, and Ψ_π values were measured using a Wescor dewpoint microvoltmeter.

Donkioporia growth, measured as radial growth rate (K_r mm day^{-1}), was optimal at -0.4 MPa and decreased sharply with decreasing water availability (Fig 19). Using four osmotica, mycelial extension of seven isolates ceased between –4.0 and –6.0 MPa when incubated at 25°C. Individual differences in the responses of isolates were observed, for example, Isolate 299.93 was less sensitive to water stress than other isolates. Apart from a slight enhancement of growth at −1.0 MPa, when using NaCl and KCl as osmotica, the general trend of response was similar for the four osmotica.

While the radial growth method is commonly used in studies on the water relations of fungi (Bruehl & Kaiser 1996), there are limitations to the method and the

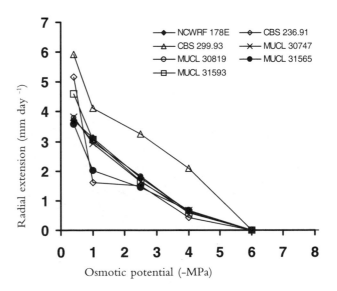

Figure 19 Radial growth of seven isolates of Donkioporia *in relation to osmotic potential (Ψ$_\pi$) at 25°C. Media was osmotically altered with glycerol. Data points represent growth rates (slopes of regression lines) for each isolate, based on three replicate measurements. Standard error of means are within the area of points.*

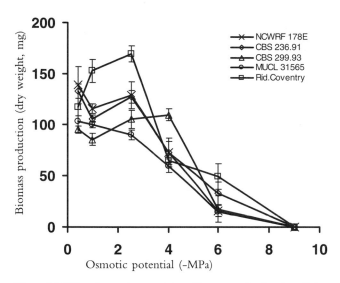

Figure 20 Effect of osmotic potential on biomass production (mycelial dry weight, mg) of five Donkioporia *isolates, grown on glycerol-amended potato dextrose broth, at 25°C. Data points, with standard error bars, are means of five replicate harvests.*

measurement of fungal biomass is considered to better reflect the growth response of a fungus. For example, growth of *Donkioporia* as assessed by measuring mycelial biomass (Fig 20) was optimal between -0.3 MPa to −2.5 MPa, a broader range than that found for radial extension (Fig 19). Otherwise the results of the previous experiment were confirmed ie *Donkioporia* does not grow at Ψ_π lower than −6.0 MPa.

In many substrata osmotic potential can be assumed to be the major element of water potential. However, in the case of woody tissue, matric potential rather than osmotic potential is likely to exert the major influence on the growth of an organism. Matric potential is thought to better reflect the problems faced by fungi in accessing water from a porous wood structure. Water availability is related to pore size and so becomes less available from within the voids of decreasing size in a woody matrix. Mindful of this, an experiment was carried out investigating the growth responses of *D. expansa* on matrically-altered media. Polyethylene glycol (PEG 20,000) was

added to a basal medium of PDB to give matric potentials in a range -0.3 to -4.0 MPa. Of the four isolates tested all grew at a matric potential of -3.5 MPa and all but one, Isolate 299.93, ceased growing at -4.0 MPa.

The results of this work on *Donkioporia* correspond well with published data for other fungi. Most wood-rotting basidiomycetes cease active growth between -5.0 and -7.0 MPa (Dix & Webster 1995). The investigation has shown that *Donkioporia* is incapable of growth at osmotic and matric potentials below *c* -7.0 MPa. For most of the isolates growth had ceased by -6.0 MPa. These values correspond to theoretical equilibrium moisture contents (EMC) of 23% to 25% in intact wood. Wood drier than this will not support mycelial extension of *D. expansa*, although if the fungus is already established it could remain viable under these and drier conditions. *Donkioporia* was found to survive for at least nine months in solute-amended media at -9.0 MPa. Some wood-rotting fungi have been known to survive in dry atmospheres (45% RH) for more than five years (Cartwright & Findlay 1958).

DONKIOPORIA CHLAMYDOSPORES

Germination and survivability

Factors influencing chlamydospore germination were investigated (Moore & Fuller 1998) and also chlamydospore survivability (Somers 1997). Procedures were developed for optimizing production, harvesting, washing and germination of chlamydospores. Chlamydospores of all isolates readily germinated on commonly used mycological media (Fig 21a). Germination also took place in various wood diffusates and on the surface of wood of different percentage moisture contents, including medieval and modern oak (Fig 21b). However, when using defined liquid media (eg Raulin's modified medium) germination was significantly reduced. No population/spore density effects on germination were observed and washing experiments confirmed that no autostimulatory or inhibitory factors were produced (Brady 1996). Chemotropic responses, eg to wood/wood extracts, have not yet been investigated.

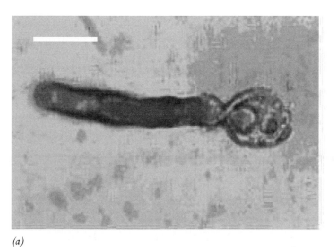

(a)

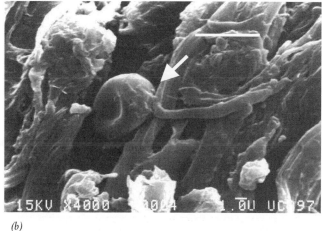

(b)

Figure 21 Germinating chlamydospores of D. expansa *on (a) malt extract agar after 24 h at 25°C and on (b) modern oak heartwood of 50% moisture content (arrowed). Bar = 10 μm.*

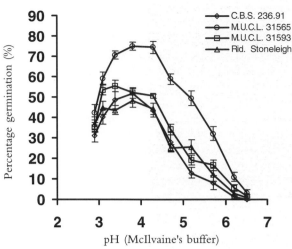

Figure 22 *The effect of pH on chlamydospore germination after 20 h and at 25°C. Potato dextrose agar was pH-adjusted using McIlvaines buffer system. Data points (with standard error bars) represent the mean percentage germination of at least one hundred spores, in each of six replicate tests.*

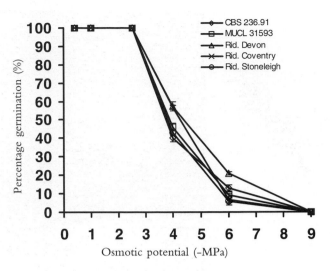

Figure 24 *The effect of osmotic potential (-MPa) on chlamydospore germination on glycerol–amended potato dextrose agar after 96 h at 25°C. Data points (with standard error bars) represent the mean percentage germination of at least one hundred spores, in each of six replicate tests.*

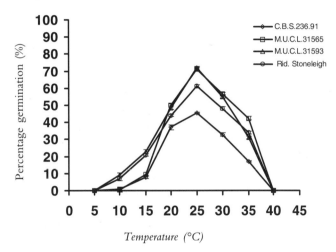

Figure 23 *The effect of temperature (°C) on chlamydospore germination after 20 h on potato dextrose agar. Data points (with standard error bars) represent the mean percentage germination of at least one hundred spores, in each of six replicate tests.*

Chlamydospores do not swell on germination; some slight shrinkage occurs as cytoplasm moves out of the spore into the extending germ–tube. Clamp connections were commonly observed on newly formed germ–tubes. Chlamydospores are long lived; tests showed that one-year-old spores readily germinated.

With regard to pH, germination was optimal in the range pH 3.1 to 4.4, with slight variation between isolates

(Fig 22). Interestingly, germ–tube extension was optimal over the broader pH range, pH 3.1 to 5.6. No germination was possible above pH 6.5 or pH 6.9, using McIlvaine's and phosphate buffers respectively, again emphasizing the acidophilic nature of *Donkioporia*.

Chlamydospore germination was markedly influenced by temperature (Fig 23): c 25°C is the optimum temperature with a min-max range extending from 5° to 40°C. Germination is possible, but extremely slow, at 5°C, reaching 14% germination after ten days. With even a moderate rise in temperature germination is greatly enhanced. At 25°C germination commenced after a lag of three hours and reached completion within twenty-four hours. Low levels of germination were possible at 38°C, but no chlamydospores germinated at 40°C.

Water is a prerequisite for the germination of *Donkioporia* chlamydospores. Percentage germination of chlamydospores decreased steadily as water potential decreased. No germination occurred at -6.0 MPa after twenty hours but germ-tubes had emerged by ninety-six hours (Fig 24). No germination was observed at -9.0 MPa.

Table 3 shows the ability of chlamydospores to germinate on wood blocks of various moisture contents. On oak wood, germination was possible at c 26–30% EMC but not at 17%. Moisture contents in the range 23–25% correspond to theoretical water potentials between -7.0 and -4.0 MPa respectively, previously shown to inhibit

Table 3 Germination of *Donkioporia* chlamydospores on woods of different equilibrium moisture contents (% EMC) after 21 days at 22°C.

Wood type	Chlamydospore growth* and wood equilibrium moisture content (% EMC)				
Modern oak	+ (115.3)	+ (26.7)	– (17.2)	– (15.2)	– (11.4)
Medieval oak	+ (173.5)	+ (29.5)	– (17.4)	– (13.4)	– (10.8)
Beech	+ (69.4)	+ (29.3)	– 17.9)	– (15.1)	– (12.1)
Pine	– (73.7)	– (23.6)	– (15.9)	– (14.2)	– (10.6)

*Chlamydospore rermination was indicated by the presence of mycelium, + and – representing mycelium present or absent Wood EMC (%) was achieved by incubating wood blocks over water and various saturated salt solutions, producing a range of relative humidities, in sealed chambers. Values are means of results obtained using ten replicate blocks.

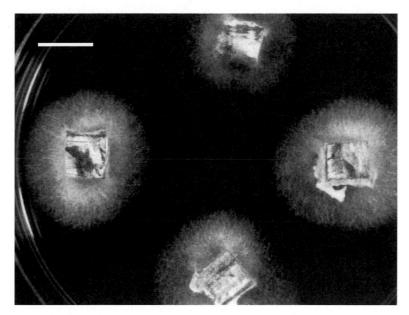

Figure 25 Regrowth of Donkioporia *mycelium (Isolate 31565) from oak wood blocks subjected to 80°C, in a dry atmosphere, for 2 hours. Bar = 10 mm.*

spore germination. The ability of chlamydospores to germinate on moist wood may indicate an important role for these propagules in the spread and establishment of *Donkioporia*.

Chlamydospore survivability was investigated by subjecting populations of chlamydospores to temperatures in the range 50° to 80°C, in both wet and dry environments. Wood blocks (8 x 8 x 6 mm) colonised by *Donkioporia* were similarly treated. When spores were subjected to temperatures of 50°, 55°, 60° and 65°C in a wet environment the extent to which germination took place was dependent on the time of exposure. When the temperature was raised to 65°C under wet conditions spores were more rapidly killed; no spores survived exposure for twenty minutes at 55°, 60° or 65°C.

Chlamydospores within *Donkioporia*-colonised wood blocks were more resistant to high temperatures because of the protective insulation afforded by the wood. Chlamydospores within wood blocks were subjected to temperatures of 50°, 55°, 60°, 70° and 80°C for periods of time of up to two hours. Three isolates survived a thirty-minute exposure at 80°C, and Isolate 31565 survived for one hour.

Spores were also exposed to temperatures of 50°, 60°, 65°, 70° and 80°C in a dry environment. Under these conditions spores were more resistant and survived two hours exposure at all temperatures to 70°C. Chlamydospore survivability in wood was even more pronounced. *Donkioporia* re-grew from oak, beech and pine wood blocks exposed for two hours at 80°C in a dry atmosphere (Fig 25).

The results show that *Donkioporia expansa* chlamydospores can survive raised temperatures for extended periods of time, particularly if protected within a substratum such as wood. Chlamydospores were also shown to have greater resistance in dry as opposed to wet environments. Kurpik and Wazny (1978) reported similar findings for mycelia of the wet-rot fungi *Gloeophyllum sepiarium* and *Coniophora puteanum*.

The above results are relevant in relation to heat treatment methods for the control of wood rot, which have been introduced in recent years (Koch et al 1989). In these methods warm air/steam is introduced into a building, usually for a period of days. In the case of *Serpula lacrymans* a temperature of at least 40°C for twenty four hours is used (Koch 1991). Having killed the active mycelium it is important to eliminate any likely sources of moisture which would make it possible for the fungus to re-establish.

In deciding protocols for heat treatments of buildings, it is important that there be awareness of the greater resistance of chlamydospores to raised temperatures, and possibly fungicides, than either mycelia or basidiospores. Chlamydospores are resistant, long-lived structures produced by *Donkioporia* and several other wood-rotting fungi. To date, chlamydospores have received inadequate attention as dispersive, survival propagules.

DEGRADATIVE ENZYMES AND WOOD DECAY

Enzymes

Donkioporia expansa is a white-rot fungus whose activities result in colonised wood having a bleached, white coloration and a stringy fibrous texture (Fig 1). Typically, white-rot fungi preferentially degrade the lignin component of wood with slower, later degradation of cellulose, hemicellulose and other cell wall components. The relative activities of ligninase, cellulase and hemicellulase enzyme systems account for the observed degradative effects. Simultaneous and sequential groups of white rot fungi are recognized, depending on the deployment patterns of these enzymes (Zabel & Morrell 1992). Campbell and Bryant (1940) characterized *Donkioporia* rot of oak wood as a simultaneous white rot.

The extracellular depolymerase enzymes of *D. expansa* were investigated in a preliminary screening experiment

Table 4 Enzymes detected in Donkioporia expansa *isolates.*

Enzyme activity	Substrate/Test
Acid protease	Casein
Acid protease	Gelatin
β-1,4 glucanase	Carboxymethylcellulose
Hemicellulase	Rutin
α -amylase	Starch
Lipase	Sorbiton monolaurate
DNAase	DNA
Pectinase	Polygalacturonic acid
Phenoloxidase	Gum guaiac
Phenoloxidase	Tannic & gallic acids
Peroxidase	Pyrogallol & peroxide
Laccase	α-naphtol
Laccase	ABTS
Tyrosinase	Cresol

(Moore 2001) using both plate detection (Hankin & Anagnostakis 1975) and spot test methods (Rayner & Boddy 1988). In the plate method, polymeric substrates (eg starch, cellulose) were incorporated into culture media of compositions appropriate for the production and activity of particular enzymes. The plates were inoculated with ten *Donkioporia* isolates and examined after seven days of incubation at 25°C. Enzyme activities were detectable by observing degradative changes in the polymer-containing media eg precipitate formation, clearance zones. In some tests the application of reagents was required in order to visualise the extent of polymer breakdown. Biochemical tests were also used which involved the application of substrates (eg phenolics) to the surface of *Donkioporia* colonies and observing for enzymatically-induced colour changes. Table 4 lists the enzymes detected in cultures of *D. expansa*. All of the enzymes listed were detected in most of the isolates.

Phenoloxidases (including laccases, tyrosinase and cathecol oxidases) are capable of modifying and degrading the complex lignin polymer. Presence of these enzymes confirms that *Donkioporia* is a white-rot fungus. Phenoloxidase enzymes have previously been reported in individual *Donkioporia* isolates by Law (1955), Domanski and Orlicz (1967) and Ritter (1983). With regard to the lignin-modifying enzymes, the results of the chemical spot tests were not always unequivocal, and variation in the complements of these enzymes was recorded in different *Donkioporia* isolates. Tyrosinase activity was not convincingly demonstrated in all isolates, which concurs with Ritter (1983) but not with Law (1955) who reported high cresolase/tyrosinase in *D. expansa* (then *Phellinus megaloporus*).

The diverse enzyme complement detected in *Donkioporia* isolates indicates a fungus that is capable of utilizing a wide range of naturally-occurring substrates. The potency of the enzyme complement is evident in the lignocellulose degradation and significant weight losses observed in oak beams in rot-affected buildings. Dry, rotten oak timbers are surprisingly light in comparison to solid oak and the soft stringy consistency of the wood offers little mechanical strength. Esser and Tas (this volume) have described changes in the chemical composition of oak wood attacked by *Donkioporia* and other decay fungi.

The production of cellulase (endoglucanase) and lignin-modifying enzymes is not surprising for a white-rot fungus, but the detection of amylase, pectinase, lipase and DNA-ase, is less usual in wood-rotting fungi. The presence of b-amylase in *Donkioporia* is interesting because there are few reports of basidiomycetes having this capability (Fogarty & Kelly 1990, Kelly pers comm). The detection of pectinase may signal a parasitic role for *Donkioporia* which has not yet been discovered. Niches in buildings now colonised by wood-rotting fungi have only existed for, at most, a few millennia but the same fungi have survived for very considerably longer in the natural environment. Boddy and Rayner (1984) found that some wood-rotting fungi in nature were already present in trees before death. Fungi causing decay in buildings (eg *Laetiporus sulphureus*, *Fistulina hepatica* and perhaps *Donkioporia*) may also already be established in timbers, as saprotrophs on stored material or as parasites in standing trees, prior to their being used for constructional purposes. With regard to wood decay caused by *Donkioporia*, Watkinson (1994) makes the point that, given its rarity (in the UK) and preference for oak, it was likely already present as mycelium or spores on timbers used in construction.

Wood decay

In a series of experiments the ability of several *Donkioporia* isolates to degrade different wood types *in vitro* was investigated (Moore 2001). The first study investigated wood decay using the CEN recommended EN113 test (EN113 1994). Following the standard procedure, heartwood blocks of oak (medieval and modern), pine and beech were placed onto MEA cultures of three *Donkioporia* isolates in rectangular jars of appropriate volume. Following incubation at 22°C for sixteen weeks, blocks were harvested and dry weights determined. Apart from beech, the mean dry weight losses of the wood blocks were insignificant. Mean dry weight losses for beech and pine were 5.4% and 1.3%, respectively. The weight losses in oak were low, at 0.6% and 2.0% for medieval and modern oak, respectively. Esser and Tas (this volume) reported similar results for oak blocks colonised by Isolate 299.93, for sixteen weeks. Various weight loss values are cited in the literature by authors who have used *Donkioporia* in decay studies. Buchwald (1986) using EN113 test procedures examined the effect of a German isolate of *Donkioporia* on nineteen wood types and reported percentage weight losses of 34–46% and 9–23% in oak sapwood and heartwood, respectively; weight losses in pine were < 3% and 27–50% in beech.

The influence of water availability on wood decomposition by *Donkioporia* was also investigated. Decay of medieval oak was assessed at five relative humidities: 100%, 97%, 93%, 85% and 75%. Pre-colonised standard wood blocks (EN113 1994), four replicates for each treatment, were exposed for sixteen weeks at 22°C in the variously humid atmospheres in rectangular flint jars (Fig 26). Significant weight losses occurred only in wood blocks incubated with *Donkioporia* in 100% RH; wood

Figure 26 Relative humidity/decay experiment showing oak wood blocks supported over water and inoculated with Donkioporia *(Isolate 31565). Bar = 15 mm.*

Table 5 Decay of medieval oak heartwood inoculated with various Donkioporia *isolates, incubated over water (100% RH) for sixteen weeks at 22°C.*

Donkioporia isolates	% Weight loss	% Moisture content
NCWRF 178E	1.7 ± 0.1	60.3 ± 2.7
CBS 236.91	3.5 ± 0.1 ★	56.9 ± 12.1
CBS 299.93	1.2 ± 0.1	58.9 ± 3.7
MUCL 30747	1.8 ± 0.6	55.2 ± 7.2
MUCL 30819	1.9 ± 0.7	53.7 ± 13.9
MUCL 31565	1.7 ± 0.7	60.2 ± 11.8
MUCL 31593	3.7 ± 1.9 ★	51.2 ± 10.4
Control	0.8 ± 0.1	35.2 ± 3.5

Values are means of four replicate blocks, ± standard deviations; ★ significant decay, compared to controls, ($P < 0.05$).

block moisture contents ranged from 51–60%. Of seven *Donkioporia* isolates tested, only two isolates, 236.91 and 31593, caused significant loss in weight, of 3.5% and 3.7%, respectively, in oak blocks (Table 5). In this experiment mycelial growth did not occur on wood blocks incubated in atmospheres below 97% RH (wood block EMC *c* 25%). This result corresponds well with the results of the solute-amended agar tests in which *Donkioporia* growth occurred at −4.0 MPa (*c* 97% RH, 25% EMC) but stopped at −6.0 MPa (*c* 95% RH, 23% EMC). Similar results have been found for other wood-rotting fungi. Freyfeld (1939) concluded that the minimum MC for growth of *Poria vaporaria* was 26% (93.6% RH) while Thedan (1941) found no weight loss in wood blocks of 24% MC (96.5% RH) exposed to *Coniophora cerebella* or *Serpula lacrymans* at *c* 94.5% RH.

The effect of direct contact of wood blocks with water was investigated in an experiment in which oak blocks of standard test size were placed on sterile moist sand. Oak blocks were pre-inoculated with *Donkioporia* Isolates 178E, 236.91 and 31565 and incubated at 22°C for sixteen weeks. Fungal growth was extensive on all blocks with some showing basidiocarp development and exudate production (Fig 27). Humidity within the chambers was high and wood blocks had moisture contents in the

range 80–90%. While weight losses, at 5%, were significant and higher than those recorded in previous decay experiments, they do not reflect the known potential of *Donkioporia* to cause extensive decay of oak timbers. However, the results do indicate a requirement for higher moisture levels and probably longer incubation periods.

In a final series of experiments an attempt was made to accelerate the decay process using miniblocks (10 x 10 x 5 mm) of oak and beech heartwood. Supported miniblocks were placed in groups of four on MEA plates, inoculated with *Donkioporia* and incubated at 22°C for sixteen weeks (Fig 28). Modern and medieval oak miniblocks incurred weight losses up to 10.5 and 7.2% respectively, depending on *Donkioporia* isolate (Table 6). Beech blocks were much less resistant to *Donkioporia* decay, showing weight losses up to 50%. The results also show the different decay capacities of *Donkioporia* isolates.

An SEM study of oak, beech and pine wood blocks colonised by *Donkioporia* isolates was carried out by Monaghan (1997). After eight weeks incubation at 20°C, by which time the blocks were covered with mycelium, blocks were sectioned in various planes and examined using a cryoscanner. Of the three wood types, oak wood was least affected by *Donkioporia*. Hyphae had ramified throughout the wood tissue but there was little evidence of white-rot attack as described by various authors, for example, Otjen and Blanchette (1986). In contrast, the beech wood attacked by *Donkioporia* isolates showed signs of lignocellulose degradation as did wood blocks colonised by *Coriolus versicolor*, included as a control. The results of the SEM study echoed the findings of the *in vitro* decay work: *Donkioporia* isolates did not significantly degrade either medieval or modern heartwood, over the time period used.

Results of decay tests indicate that lengthy incubation periods are required for significant decay of oak heartwood. In addition, moisture requirements for growth and decay by *Donkioporia* are high. Mycelial growth was not observed on wood blocks of less than *c* 27% moisture

Figure 27 Donkioporia *(Isolate 31565) colonized oak heartwood on moist sand showing exudate production (arrowed). Bar = 10 mm.*

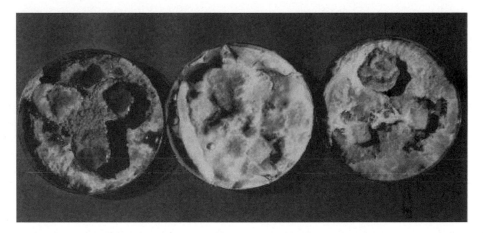

Figure 28 Accelerated decay test showing miniblocks of oak heartwood inoculated with Donkioporia.
(L-R) Isolates 31565, 236.91 and 178E supported over malt extract agar (MEA). 90 mm Petri dish.

content and for decay, wood moisture contents of greater than 50% were required. It is well established in the literature that heartwood is more resistant to decay organisms than sapwood. Cartwright and Findlay (1958) recorded dry weight losses of 35% in oak sapwood inoculated with *Donkioporia* and < 10% in heartwood while Ritter (1988) measured 10% weight loss in oak (heartwood?) after four months at 20°C. Szabo and Varga (1995) reported mean weight losses of 7.8%, using Scots pine sapwood blocks, colonised by a Hungarian *Donkioporia* isolate. Weight losses reported by Buchwald (1986) for *Donkioporia* inoculated oak heartwood are of

the same order to those measured in the decay tests reported in the present study (ie 5–10%).

CONTROL OF *DONKIOPORIA*

Growth of *Donkioporia* in buildings can be controlled by eliminating problematic sources of moisture, which facilitate activity of this wet-rot fungus. Badly rotten, mechanically weakened timbers should be replaced or variously repaired. Affected areas should be dried out and subsequently maintained in a dry condition, ideally monitored, such that wood moisture levels do not exceed 20%.

Table 6 Decay of heartwood miniblocks by Donkioporia expansa *after sixteen weeks at 22°C.*

Wood type	*Mean % weight losses	Mean % moisture contents
Modern oak	5.3; 10.5; 2.6; (0.1)	111.6; 132.6; 108.7; (76.4)
Medieval oak	7.2; 3.2; 3.5; (1.5)	169.3; 145.5; 142.5; (134.2)
Beech	24.8; 16.1; 50.8; (0.1)	129.4; 110.3; 101.5; (63.5)

* Values are mean weight losses of ten replicate blocks (10 x 10 x 5 mm), inoculated with (L–R), Isolates 178E, 236.91 and 31565. Bracketed values are for uninoculated control blocks. Weight losses for inoculated blocks were significantly greater when compared to control blocks, ($P < 0.05$).

The present study has shown that while *Donkioporia* may survive under these conditions, neither mycelium nor chlamydospores can grow at moisture levels less than *c* 25%. Kleist and Seehann (1999) maintain that *Donkioporia* cannot spread to adjacent dry timbers or grow through brickwork.

In some situations the use of additional control measures may be necessary (Ridout this volume). For example, if the drying-out period is likely to be prolonged, use of a fungicide might be essential to prevent regrowth of *Donkioporia*. In the light of experience and the undesirable legacy following from the use of noxious pesticides and preservatives in the past, it is not in the spirit of the Woodcare project to advocate the use of chemicals likely to be hazardous to health, damaging to the fabric of buildings or to the environment. During the last decade boron-based fungicides have come to the fore, with claims of their being more acceptable from health and environmental perspectives. While there are likely to be hazards associated with the use of any biocide, there are circumstances in which the use of boron fungicides might be acceptable. Some boron preparations are applied as colourless pastes onto timber surfaces, from where they diffuse into the wood structure. As moisture is essential for adequate penetration by these fungicides, one could envisage their being used in situations where timbers were slow to dry out.

Three boron-containing fungicides (B40, Biokil and Deepwood 50) were assessed for their effectiveness against three *Donkioporia* isolates (31565, 299.93 and Rid. Stoneleigh) using an agar plate diffusion method. Fungicides B40 and Deepwood 50 totally inhibited growth of the three *Donkioporia* isolates. Restricted growth occurred in the presence of Biokil. Inoculum plugs from the B40 and Deepwood plates were assessed for viability by subsequently transferring them to fresh PDA. Growth re-established from the B40-exposed plugs but not from those exposed to Deepwood 50, indicating fungistatic and fungicidal effects, respectively. The results of this preliminary test indicated that the boron fungicide Deepwood 50 shows promise in controlling growth of *Donkioporia* and warrants further assessment using standard procedures. However, Hegarty and Buchwald (1988) found that *Donkioporia* basidiospores germinated successfully on various wood blocks, including oak, treated with sodium polyborate (Basilit B at 2 Kg/m³). They also noted that this level of fungicide was twice that needed to prevent mycelial attack.

SUMMARY AND CONCLUSIONS

Oak timbers, often in historic buildings, are subject to serious decay by the wet-rot fungus, *Donkioporia expansa*. The possibility that its presence in oak wood might also facilitate deathwatch beetle attack, and further damage to timbers, adds to the relevance of this previously disregarded fungus. Other hardwoods and some softwoods are also susceptible to *Donkioporia*. A review of *Donkioporia* literature revealed that while there is ample taxonomic and morphological detail available on the fungus, virtually nothing is known of its biology.

A collection of twelve *Donkioporia expansa* isolates was assembled and maintained. Nutritional and environmental factors influencing growth *in vitro* were investigated and requirements for bulk culture of the fungus were established. Excellent mycelial growth was obtained in malt extract, potato dextrose and Raulin's modified media; optimal growth was obtained in the range 25°–30°C and at pH 4.1–4.4. Growth kinetics were determined for each isolate; radial growth rates for most isolates were *c* 4–5 mm day⁻¹, at 25°C. Cultural conditions were adapted to maximise mycelial biomass yields.

The availability of a *Donkioporia* collection afforded the opportunity to compare colony morphologies and other characteristics. Features of basidiomes differentiating *in vitro* were also examined and quantified. Considerable variation between isolates was observed and this was reflected in different species codes. As basidiospores were unavailable, monokaryon crosses could not be used to examine the genetic relatedness of the isolates. Hyphal interactions between isolates were examined but did not yield useful information. The observed variation prompted further study of the relationships between isolates, using DNA analyses. RAPD analyses indicated four subpopulations of *Donkioporia* in the small collection of isolates available.

The water relations of *Donkioporia* isolates were investigated under various experimental conditions, using a variety of inorganic and polyol osmotica. The results show that *Donkioporia* is incapable of growing at water potentials less than -6.0 MPa, which corresponds to a theoretical wood moisture content of *c* 25%. However, mycelia and spores can survive drier conditions for prolonged periods of time. The influence of matric potential on growth was also examined, using PEG 20,000-amended culture media. No isolate was capable of growing at matric potentials below *c* -4.0 MPa. The results of this work correspond well with published data on the water requirements of other wood-rotting basidiomycetes.

A wide range of extracellular enzymes was detected in *Donkioporia* isolates, including phenoloxidases which confirmed the white-rot status of the fungus. Not all isolates produced tyrosinase. The significance of other depolymerases was considered in relation to the possible natural niche of *Donkioporia*.

Various wood decay experiments were undertaken using different wood types and *Donkioporia* isolates. Significant decay, indicated by weight losses of 5–10% in oak heartwood blocks, was observed only when blocks were incubated in wet conditions (100% RH/wet sand). SEM examination of eight-week-colonised oak blocks showed only limited white-rot effects. Results indicated that substantial decay of oak heartwood by *Donkioporia* occurs only after prolonged exposure to high moisture levels.

Chlamydospores are a feature of *Donkioporia* cultures, on which they form in very large numbers within days. Chlamydospores are thick-walled, resistant propagules likely to play a role in the survival and dispersal of the fungus. Chlamydospores germinated on most media within hours of incubation. Germination rates were optimal at 25°C and pH 3.1 to 4.4. Germination was extremely slow at 5°C and was inhibited at 40°C. Chlamydospores are long-lived; one-year-old spores readily germinated. Exposure to elevated temperatures also showed them to be resistant propagules. This has implications should heat treatment of *Donkioporia*-affected buildings be proposed as a control measure. In relation to control, *Donkioporia* was shown to be sensitive to boron-based fungicides, which might be an acceptable chemical option in situations where environmental control of fungal growth is not immediately effective.

In addition to facilitating work on the extraction of chemicals from *Donkioporia*, the work has contributed to knowledge of a previously little-researched wood-rotting fungus. A comprehensive knowledge of the biology of *Donkioporia* will be essential to understanding the association between this fungus and deathwatch beetle and also in the development of strategies to control fungus and beetle.

BIBLIOGRAPHY

Bassett C, Sherwood R T, Kepler J A and Hamilton P B, 1967 Production and biological activity of fomannosin, a toxic sesquiterpene metabolite of *Fomes annosus*, in *Phytopathology*, **57**, 1046–1052.

Baxter D V, 1950 Some resupinate polypores from the region of the Great Lakes, in *XX. Paper Michigan Academy of Sciences*, **34**:1, 41–56.

Boddy L and Rayner A D M, 1984 Fungi inhabiting oak twigs before and at fall, in *Transactions of the British Mycological Society*, **82**, 501–505.

Bondartsev A.S, 1971 Polyporaceae of the European USSR and Caucasia, in *Israel Program for Scientific Translations*, Jeruselam, 896 S.

Bourdot H and Galzin A, 1928 *Hymenomycetes de France*, I. Paris, 685–686.

Brady K, 1996 *An Examination of the Factors affecting Chlamydospore Germination in Donkioporia expansa*, unpublished BSc thesis, Department of Botany, University College Dublin.

Breitenbach J and Kränzlin F, 1986 *Fungi of Switzerland, 2: non gilled fungi*, Lucerne, Verlag Mykologia.

Bruehl G W and Kaiser W J, 1996 Some effects of osmotic water potential upon endophytic *Acremonium* spp. in culture, in *Mycologia*, **88**, 809–815.

Buchwald G, 1986 On *Donkioporia expansa* (Desm.) Kotl. & Pouzar, in *IRG Document no.* IRG/WP/1285, Stockholm.

Butler G M, 1995 Induction of precocious fruiting by the diffusable factor in the polypore *Phellinus contiguus*, in *Mycological Research*, **99**:3, 325–329.

Campbell W G and Bryant S A, 1940 A chemical study of the bearing of decay by *Phellinus cryptarum* Karst. and other fungi on the destruction of wood by the deathwatch beetle (*Xestobium rufovillosum* De G.), in *Biochemical Journal*, **34**,1404–1414.

Cartwright K St G and Findlay W P K, 1936 *The Principal Rots of English Oak*, London, HMSO.

Cartwright K St G and Findlay W P K, 1958 *Decay of Timber and its Prevention*, 2nd edition, London, HMSO.

Cymorek S and Hegarty B, 1986 A technique for fructification and basidiospore production by *Serpula lacrymans* (Schum. Ex Fr.) St Gray in artificial culture, in *IRG Document no.* IRG/WP/2255, Stockholm.

Decock C and Hennebert G L, 1994 Wood-decaying fungi in Belgian buildings: Four years of investigations, in *Poster Abstracts, Fifth International Mycological Congress, Vancouver*.

Dix N and Webster J, 1995 *Fungal Ecology*, London, Chapman and Hall.

Domanski S and Orlizc A, 1967 *Polyporus megaloporus* Pers. in the family Polyporaceae s.str., in *Acta Mycologica* **3**, 51–62.

Domanski S, 1972 *Fungi. Polyporaceae I., Mucronoporaceae I. (resupinate)*, in Flora Polska (Grzyby),Warsaw and Springrield Va. (The original Polish work of 1965 edited in English), 145–146.

Donnelly D M X, 1997 *Biologically active compounds from Donkioporia expansa*, Final Report, EU Woodcare Research Project, EG-BKK.

Dörtfelt H and Sommer B, 1973 *Poria expansa* (Desm.) H. Jahn im Botanischen Garten Halle gefunden, in *Mykologisches Mitteilungsblatt* **17**, 44–47.

Ellis M B and Ellis P, 1990 *Fungi Without Gills*, London, Chapman & Hall.

EN113, 1994 Method of testing for determining the protective effectiveness of wood preservatives against wood destroying basidiomycetes, in *European Committee for Standardisation*, CEN/TC 38 N901.

Fisher R C, 1940 Studies of the biology of the deathwatch beetle, *Xestobium rufovillosum* De G. III. Fungal decay in timber in relation to the occurrence and rate of development of the insect, in *Annals of Applied Biology*, **27**, 545–557.

Fogarty W M and Kelly C P, 1990 Recent advances in microbial amylases, in *Microbial enzymes in Biotechnology*, (eds) Fogarty W M and Kelly C P, 2nd Edition, Elsevier Press.

Freyfeld E E, 1939 The effect of humidity on the growth of wood-destroying fungi in timber, in *Sovetsk. Bot*, **1**, 99–103: Abstract in *Reviews of Applied Mycology*, **18**, 644.

Gilbertson R L and Ryvarden L, 1986 *North American Polypores*, Oslo, Fungiflora.

Green M P, 1998 *RAPD analysis of* Donkioporia expansa *Desmaz*, unpublished BSc thesis, Department of Botany, University College Dublin.

Guillitte O, 1992 Epidemiologie des attaques, in *La merule et autres champignons nuisibles dans les batiments*, Jardin Botanique National de Belgique, Meise, 2ème édition, 34–42.

Hankin L and Anagnostakis S L, 1975 The use of solid media for detection of enzyme production by fungi, in *Mycologia*, **67**, 597–607.

Harris R F, 1981 Effect of water potential upon microbial growth and activity, in *Water Potential Relations in Soil Microbiology*, (eds) Parr J F, Gardner W R and Elliott L F, Soil Science Society of America, Special Publication **9**, Madison, Wisconsin, Soil Science Society of America, 23–84.

Hegarty B and Buchwald G, 1988 The influence of timber species and preservative treatment on spore germination of some wood-destroying basidiomycetes, in *IRG Document No.:* IRG/WP/2300, Stockholm.

Heim R, 1942 Les champignons destructeurs de bois dans les habitations, in *Institut Technologie du Batiment et des Travaux Publiq*, Circulaire Series II, **1**, 1–27.

Jahn H, 1967 Die resupinaten Phellinus-Arten in Mitteleuropa, mit Hinweisen auf die resupinaten Inonotus –Arten und *Poria expansa* Desm. = *Polyporus megaloporus* Pers, in *Westfälische Pilzbriefe*, **6**, 37–124.

Jahn H, 1971 Resupinate Porlinge, *Poria* s. *lato*, in Westfalen und im nordlichen Deutchland, in *Westfälische Pilzbriefe*, **8**, 41–68.

Julich W, 1984 *Die Nightblatterpilze, Gallerpilze und Bauchpilze*, Stuttgart – New York, Gustav Fischer Verl.

Kleist G and Seehann G, 1999 Der Eichenporling, *Donkioporia expansa* – ein wenig bekannter Holzzerstörer in Gebäuden, in Zeitschrift Für Mykologie, Band 65/1: 23–32.

Koch A P, Kjerulf-Jenson C and Madsen B, 1989 New experiences with dry rot in Danish buildings, heat treatment and viability, in *IRG Document No.*: IRG/WP/1423, Stockholm.

Koch A P, 1991 The current status of dry rot in Denmark, in *Serpula lacrymans: Fundamental Biology and Control Strategies*, (eds) Jennings D H and Bravery A F, Chichester, Wiley.

Kotlaba F and Pouzar Z, 1973 Donkioporia Kotl. Et Pouz., a new genus for *Poria megalopora* (Pers.) Cooke, in *Persoonia*, **7**, 213–216.

Kotalba F, 1984 *Zemepisne rozsireni a ekologie chorosu (Polyporales s.l.) v Ceskoslovensku*, Praha.

Kurpik W and Wazny J, 1978 Lethal temperatures for the wood-destroying fungi *Coniophora puteana* Fr. and *Gloeophyllum sepiarium* (Wulf.) Karst, in *Material und Organismen*, **13**:1, 1–12.

Law K, 1955 Laccase and tyrosinase in some wood-rotting fungi, in *Annals of Botany*, N.S., **XIX**:76, 561–570.

Mangin L and Patouillard N, 1922 Sur la destruction de charpantes au chateau de Versailles par le *Phellinus cryptarum* Karst. in *Compte Rendue Academie des Sciences* **T 175**:9, 389–394.

Monaghan S, 1997 *A Scanning Electron Microscopy study of the effects of Donkioporia expansa, a White Rot Basidiomycete on Four Wood Types*, unpublished BSc thesis, Department of Botany, University College Dublin.

Moore C, 2001 *Morphological and Physiological Studies on the White-Rot Fungus Donkioporia expansa*, unpublished PhD thesis, University College Dublin.

Moore C and Fuller H, 1997 Water relations of the oak rot fungus, *Donkioporia expansa*, in *Proceedings of the Irish Botanists Meeting*, University College Galway, 64.

Moore C and Fuller H, 1998 Influence of osmotic potential on mycelial growth and chlamydospore germination of *Donkioporia expansa*, Proceedings of the Irish Botanists Meeting, University College Dublin, in *Biology and Environment*, **98B**: 60.

Moreth U and Schmidt O, 2000 Identification of indoor rot fungi by taxon-specific priming polymerase chain reaction, in *Holzforschung* **54**:1, 1–8.

MUCL, 1998 *Catalogue of Mycotheque de L'Universitie Catholique de Louvain-la-Neuve*, Louvain, Belgium.

Nakasone K K, 1990 *Cultural Studies and Identification of Wood-inhabiting Corticiaceae and Selected Hymenomycetes from North America*, in Mycologia Memoir **15**, J. Cramer, Berlin-Stuggart.

Nobles M K, 1965 Identification of cultures of wood-inhabiting Hymenomycetes, in *Canadian Journal of Botany*, **43**, 1097–1139.

O'Shea C M, 1996 *A Study of the Morphological and Growth Characteristics of the Oak-Rotting Fungus Donkioporia expansa*, unpublished BSc thesis, Department of Botany, University College Dublin.

Otjen L and Blanchette R A, 1986 A discussion of microstructural changes in wood during decomposition by white rot basidiomycetes, in *Canadian Journal of Botany*, **64**, 905–911.

Peters J, 1996 *Genetic Analysis of Donkioporia expansa, a Wet-Rot Fungus*. Unpublished MSc thesis, Department of Botany, University College Dublin.

Rayner A D M and Boddy L, 1988 *Fungal Decomposition of Wood, its Biology and Ecology*, Chichester, John Wiley & Sons.

Ridout B, 2000 *Timber Decay in Buildings: The Conservation Approach to Treatment*, London, E & FN Spon.

Ritter G, 1983 Nuefund von *Donkioporia expansa*, in *Boletus*. Berlin, **7**:1, 3–4.

Ritter G, 1988 Verstärktes Auftreten des Ausgebreiteten Hausporlings, in *Holztechnologie*, **29**:5, 226–228.

Ritter G, 1992 Mykofloristische Mitteilung VII. Zur Verbreitung von *Donkioporia expansa* in den östlichen Bundesländern, in *Boletus* **16**:1, 26–28.

Robinson R A and Stokes R H, 1955 *Electrolyte Solutions*, New York, Academic Press.

Ryvarden L and Gilbertson R L, 1993 *European Polypores*, Part 1. Synopsis Fungorum 6, Oslo, Fungiflora.

Sharland P and Rayner A D M, 1986 Mycelial interactions in *Daldinia concentrica*, in *Transactions British Mycological Society*, **86**:4, 643–649.

Somers J, 1997 *Further investigation of the factors affecting chlamydospore germination and survivability in Donkioporia expansa*, unpublished BSc thesis, Department of Botany, University College Dublin.

Stalpers J A, 1979 Identification of wood-inhabiting Aphyllophorales in pure culture, in *Studies in Mycology*, **16**: C.B.S.

Szabo I and Varga F, 1995 Additional data on the occurrence and biology of *Donkioporia expansa* (Desmaz.) Kotl. Et Pouz, in *Mikologiai Kozlemenyek*, **34**:1, 53–58.

Thedan G, 1941 Untersuchungen Über die Feuchtigsheitanspruche der wichtigsten in Gebäuden auftretenden holzzerstörenden Pilze, in *Agnewandte Botanik*, **23**: 189–252.

Van Acker J and Stevens M, 1996 Laboratory culturing and decay testing with *Physisporinus vitreus* and *Donkioporia expansa* originating from identical cooling tower environments show major differences, in IRG Document no. IRG/WP/96-10184,8s, Stockholm.

Watkinson S, 1994 The physiology and morphology of fungal decay in buildings, in *Building Mycology*, (ed) Singh J, London, E. & F.N. Spon, Chapter 4.

Wazny J and Czajnik M, 1963 Zum auftreten holzzerstörender pilze in Gebäuden in Polen, in *Folia Forestalia Polanica*, ser. B, 5–17.

White N A, Palfreyman J W and Staines H J, 1996 Fungal colonisation of a nineteenth century wooden frigate, in *Material und Organismen*, **30**:2, 117–131.

Zabel R A and Morrell J J, 1992 *Wood Microbiology, Decay and its Prevention*, New York, Academic Press.

MATERIALS AND EQUIPMENT

HR 33T Dew Point Microvoltmeter

Wescor, inc., 459 South Main Street, Logan, Utah 84321, USA.

Deepwood 50, boron wood preservation

Safeguard Chemicals Limited, Unit 6, Redkiln Close, Redkiln Way, Horsham, Sussex RH13 5QL, UK.

B40 and Bio-Kil, boron pastes

Bio-Kil Chemicals Limited, Brickyard Industrial Estate, New Road, Gillingham, Dorset SP8 4BR, UK.

ACKNOWLEDGEMENTS

Professor D M X Donnelly, Department of Chemistry, University College Dublin, provided the opportunity and financial support to Hubert Fuller to participate in the EU Woodcare Project. Research facilities in the Department of Botany were kindly made available by Professor Martin Steer. Ridout Associates (Stourbridge, UK) provided medieval oak samples, *Donkioporia* basidiocarps and also arranged for a visit to Stoneleigh Abbey, Warwickshire. Dr Ferenc Varga, Sopron, Hungary, kindly donated Isolate Var 110a.

The following students at University College Dublin assisted through their undertaking postgraduate, undergraduate and bursary projects on various aspects of *Donkioporia*: Killian Brady BSc, Coleman Burgess BSc,

Katharine Fuller MSc, Michael Green BSc, Suzanne Monaghan BSc, Ciara O'Shea BSc, Jean Peters MSc, Michael Rice BSc, Joanne Sommers BSc and Sinead Weldon BSc. Technical support was provided by John Moran and Dermot McKeon.

The DNA analysis would not have been possible without the generous expert assistance of colleagues Professor Matthew Harmey and Dr Tommy Gallagher. Colm Moore's postgraduate studies on *Donkioporia* were financially supported by a research scholarship (Forbairt/ Enterprise Ireland) and a research demonstratorship (University College Dublin). Both awards are gratefully acknowledged.

AUTHOR BIOGRAPHIES

Colm Moore is a postgraduate student in the Department of Botany, University College Dublin, where he is completing a doctoral thesis on the wood-rotting fungus *Donkioporia expansa*.

Hubert Fuller is a mycologist who has wide research experience of parasitic and saprotrophic fungi. He is a lecturer in the Department of Botany, University College Dublin. For many years a member of the British Mycological Society, Dr Fuller participated in the EU Woodcare Project in collaboration with Professor D M X Donnelly, University College Dublin, to investigate aspects of the biology and chemistry of *Donkioporia expansa*. He is currently involved in research on another wood-rotting fungus, *Schizophyllum commune*.

The roles of location, age and fungal decay in the chemical composition of oak

Petra Esser★ and Albert Tas

TNO Building and Construction Research, PO Box 49, 2600 AA Delft, The Netherlands;
Tel: + 31 15 284 2000; Fax: + 31 15 284 3990

Abstract

The chemical composition of oak and Scots pine was studied in relation to the location, age and rate of decay present in wood. Pyrolysis mass spectrometry was found to be the most suitable method for detection of early decay in wood samples from historic buildings.

KEY WORDS

Oak, monitoring, location, age, fungal decay.

INTRODUCTION

The determination of the extent of decay in old timber structures is a complex task. This determination is often done by visual inspection of the surface of wooden parts which may lead to subjective judgements. Clearly, a more objective criterion for the degree of wood decay is needed. This would enable a more selective replacement of the affected parts in timber constructions.

During decay the chemical composition of wood may change, for example, the ratio between polysaccharides and lignins. Therefore, monitoring the degree of decay can be based on the chemical composition of wood parts. Because structures need to be preserved intact as much as possible, larger wood samples are often not available for investigating chemical composition. Applied chemical methods should be effective in handling small samples from several places in the structure without harming its construction. Moreover, the method of analysis should provide a fast analytical report on the wood's condition during reconstruction work.

Pyrolysis-direct chemical ionization mass spectrometry (Py-DCI/MS) is a recently developed technique for the improved direct analysis of complex samples. The method allows rapid analysis of minor sample amounts of only 50 g to 1 mg, providing a more detailed insight into the original chemical composition of the samples. For example, the ratio between cellulose, hemicellulose and lignin can be expected to be reflected in the pyrolysis patterns obtained by this new pyrolysis approach.

Principal components-discriminant analysis (PC-DA), a multivariate data analysis (MVA) technique and display technique, is very effective in the evaluation of these complex patterns which cannot be judged by eye only.

CHANGES IN TIMBER CHEMISTRY WITH LOCATION, AGE AND DECAY

Py-DCI/MS of timber samples from buildings

Wood samples (37 in total) were collected from seven historic buildings also used for monitoring conditions and acoustic detection (Van Staalduinen et al, this volume). In the set of wood samples tested for the degree of decay, several potential sources of variation were present:

- (G) sound oak wood
- (A1) affected by fungus
- (A2) affected by fungus and beetle
- (A3) heavily affected by beetle.

This selection was to a certain extent arbitrary and samples were inhomogeneous. In Table 1 an overview of the origin of the wood samples is given, including two new sound oak reference samples. The code numbers of the wood samples correspond to the code numbers in Figures 1 to 4.

During evaluation of the Py-DCI/MS spectra it became apparent that some of these factors contributed to the variation between the samples. An important source of variation was observed in samples from different origins (see Fig 1). For example, samples from Goes, Trippenhuis, Berlikum and Bishopstone were found to reflect major differences according to their locations. Differences in the chemical composition of the timber due to the origin of the wood seems the most logical explanation.

For Franeker, Goes and Middelburg the number of samples from each building was large enough (see Table 1 and Fig 2) to identify the age of the building as a cause of differentiation. However, origin may still be the cause of the chemical differences observed. From these results it was concluded that the determination of the degree of decay has to be carried out by location, using locally present sound wood as an internal reference. The differences in age (felling dates 1475–1650) from the wood sampled may have been too small in this case.

Subsequent data analysis by location was carried out only when sufficient samples were present. Analysis of samples from Sloten, Goes, Franeker and Trippenhuis revealed that only a partial separation between sound and

★ Author for correspondence

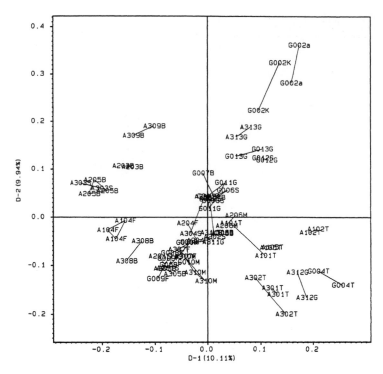

Figure 1 Discriminant plot of Py-DCI/MS spectra of 37 oak samples from seven buildings (see Table 1).

Table 1 Selected wood samples from buildings for chemical analysis in the WOODCARE project.

TNO code	Date	Origin of samples		Quantity (g)	Origin of the samples
SOUND EUROPEAN OAK					
G.01K	1995	NL	V.d. Berg A	>> 50.00	Sound reference from trader, heartwood
G.02K	1995	NL	V.d. Berg B	>> 50.00	Sound reference from trader, heartwood
G.03S	1970	NL	Sloten-6	12.00	Church tower, beam, renovated in 1970
G.04T	1650	NL	Trippenhuis-1	5.80	Room number 2.09, beam 27, next to site of endoscopy
G.05S	1600	NL	Sloten-2	8.70	Church tower, same beam as Sloten 1 and 3
G.06S	1600	NL	Sloten-5	11.70	Churchtower, same beam as Sloten 4
G.07B	1778	NL	Berlikum-3	33.00	Church, roof structure, beam
G.08F	1475	NL	Franeker-1	14.20	Church, roof structure, behind the organ
G.09F	1475	NL	Franeker-2	12.90	Church tower, beam
G.010M	1650	NL	Middelburg-1	17.50	Church roof, same beam as Middelburg 2 and 3
G.011G	1620	NL	Goes-1	56.90	Church roof, same beam as Goes 4
G.012G	1620	NL	Goes-2	96.90	Church roof, same beam as Goes 5
G.013G	1620	NL	Goes-3	39.40	Churchroof, same beam as Goes 6
DECAYED EUROPEAN OAK					
category A1: fungal decay only					
A1.01T	1650	NL	Trippenhuis-2	4.60	Room 'Buis' 2.36, beam 9, connected to corridor, heartrot?
A1.02T	1650	NL	Trippenhuis-4t/m6	16.70	Attic, beam 56 , fungal decay?
A1.03S	1600	NL	Sloten-7		Church roof, truss part
A1.04F	1475	NL	Franeker-3	25.30	Church tower, beam, fungus and *Anobium*
A1.05T	1650	NL	Trippenhuis-3	4.60	Hall, beam in facade, sound/fungal decay?
category A2: fungal decay and deathwatch beetle larvae					
A2.01S	1600	NL	Sloten-1	16.00	Church tower, same beam as Sloten 2 and 3
A2.02S	1600	NL	Sloten-9	130.00	Church, beam, roof structure
A2.03B	1778	NL	Berlikum-1	14.90	Church, beam, roof structure
A2.04F	1475	NL	Franeker-4	32.60	Church tower, beam, *Xestobium* / fungal decay?
A2.05B	1325	UK	Bishopstone-2	110.00	Church, old roof, *Xestobium* and fungal decay
A2.06M	1650	NL	Middelburg-2	41.45	Church roof, same beam as Middelburg 1 and 3
category A3: heavy deathwatch beetle larvae, bore dust of larvae					
A3.1T	1650	NL	Trippenhuis-7	3.20	Attic, beam 68, *Xestobium* /pellets
A3.2T	1650	NL	Trippenhuis-8	5.70	Attic, beam 71, light *Xestobium* decay
A3.3S	1600	NL	Sloten-3	17.00	Church tower, same beam as Sloten 1 and 3
A3.4S	1600	NL	Sloten-4	15.00	Church tower, same beam as Sloten 5
A3.5B	1778	NL	Berlikum-2	76.80	Church, beam, roof structure
A3.6B	1778	NL	Berlikum-4	108.00	Church roof, bore dust *Xestobium* from decayed beam
A3.7F	1475	NL	Franeker-5	22.00	Church, middle part, beam
A3.8B	1325	UK	Bishopstone-1	300.00	Church, flooring with *Xestobium*
A3.9B	1325	UK	Bishopstone-3	134.00	Church, old roof, heavy *Xestobium* (bore dust/fungus?)
A3.10M	1650	NL	Middelburg-3	26.40	Church roof, same beam as Middelburg 1 and 2
A3.11G	1620	NL	Goes-4	28.00	Church roof, same beam as Goes 1
A3.12G	1620	NL	Goes-5	69.90	Church roof, same beam as Goes 2
A3.13G	1620	NL	Goes-6	41.00	Church roof, same beam as Goes 3

Note: The wood samples were classified as sound to decayed (A1–A3) based on visual assessments

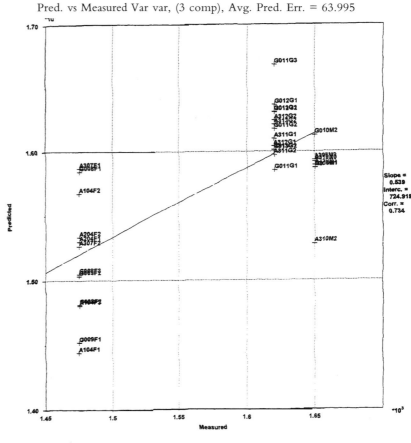

Figure 2 *PLS regression plot of the correlation between age and pyrolysis mass spectrometry profiles of oak from Goes (1620), Franeker (1475) and Middelburg (1650).*

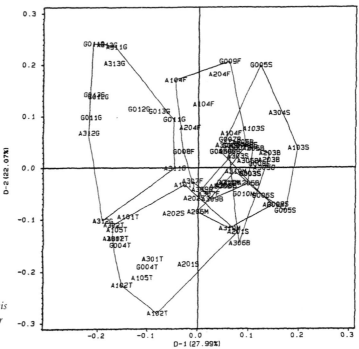

Figure 3 *Discriminant plot of pyrolysis mass spectrometry spectra of the locations Berlikum, Goes, Franeker, Trippenhuis and Sloten. The locations from Frysland (Berlikum, Franeker and Sloten) are more related: this may be due to similar oak sources used in construction.*

affected samples was possible (see Figs 3 and 4). However, it has to be noted that several types of affected sample were present in relatively small data sets.

Conclusion

Py-DCI/MS is a powerful approach for the determination of wood decay. Small quantities of samples sufficed and larger numbers of samples can be handled in a short time. Calibration of the method needs to be improved by

taking sufficient well-defined samples for each location. Additional experiments would be necessary to study the effect of age, as the number of samples was not large enough to draw conclusions on this aspect.

CHANGES IN COMPOSITION OF TIMBER DURING FUNGAL DECAY

Model experiments were performed on sound European oak heartwood samples which were subjected to fungal

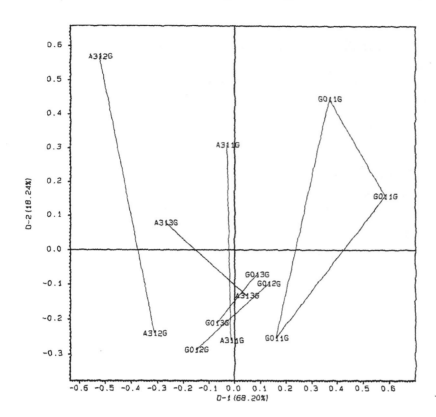

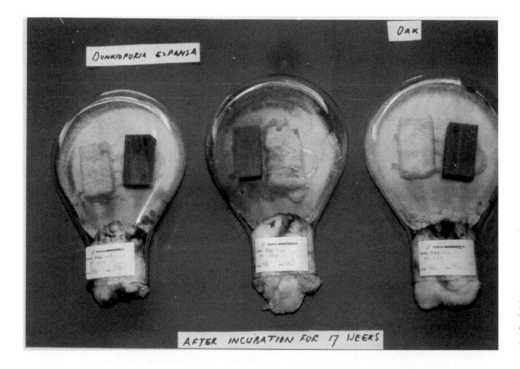

Figure 4 Discriminant plot of pyrolysis mass spectrometry spectra of sound oak (code G011 to G012) and affected oak (code A3011 to A3013) from Goes.

Figure 5 Exposure of oak to Donkioporia expansa *on malt agar in Kolle flasks, according to EN 113.*

decay under sterile conditions. Culturing conditions were identical to those described in the European standard for basidiomycetes testing on wood (Kolle flask method, cf. EN 113, 1996). The selection of fungi was based on those mentioned in the literature (eg Fisher 1940, Fisher 1941) as being associated with deathwatch beetle attack. The fungi *Coniophora puteana* (Cp), *Donkioporia expansa* (De) and *Fistulina hepatica* (Fh) and a mixture of *Fistulina hepatica* and *Donkioporia expansa* (FhDe) were applied. In addition, Scots pine heartwood and sapwood samples were subjected to fungal attack

with *Coniophora puteana* and *Donkioporia expansa*. The period of fungal attack varied from seven to ten and 16–20 weeks respectively (Fig 5).

The mass loss during these periods in European oak was always insignificant (less than 3%). Beech reference samples showed mass losses of 15% after four weeks, and up to 37% after 16 weeks. The mass loss in Scots pine heartwood was less than 3%, and 25–30% in the Scots pine sapwood samples.

The aim of the experiments was to monitor changes in chemical composition induced by fungal attack on wood

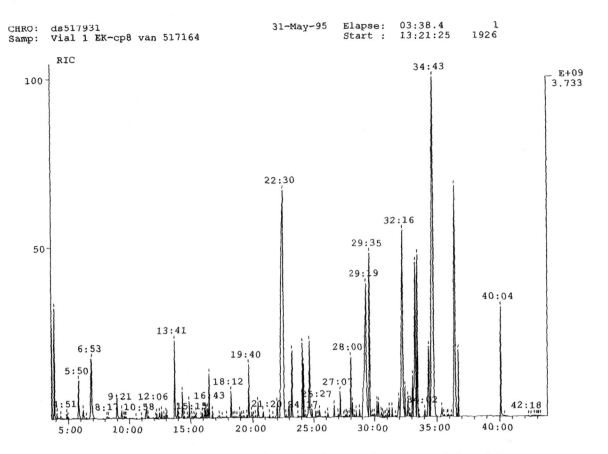

CHRO: ds517931   31-May-95   Elapse: 03:38.4   1
Samp: Vial 1 EK-cp8 van 517164   Start :   13:21:25   1926

Figure 6 GC-MS chromatogram of L/N extract of affected oak by Coniophora puteana *(eight weeks).*

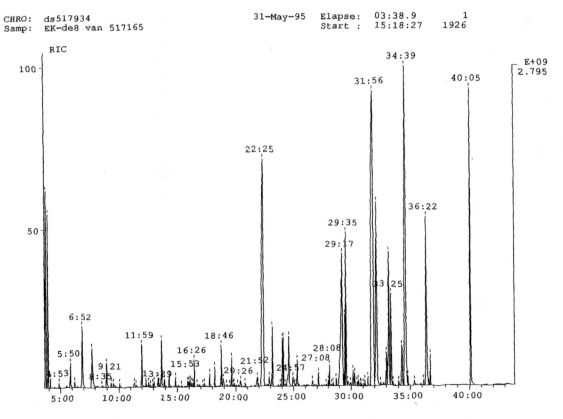

CHRO: ds517934   31-May-95   Elapse: 03:38.9   1
Samp: EK-de8 van 517165   Start :   15:18:27   1926

Figure 7 GC-MS chromatogram of L/N extract of affected oak by Donkioporia expansa *(eight weeks).*

Table 2 *L/N extracts of affected oak wood (*Coniophora puteana *and* Donkioporia expansa *after 8 weeks).*

Time	Compound	Ek-Cp8	Ek-De8
4:51	toluene	+	+
5:50	hexanal	+	+
6:12	furfural	+	+
6:53	furfural	+	+
7:42	unknown (98, 100)	–	+
8:55	cyclohexanon	+	+
9:21	heptanal	+	+
9:38	acetylfuran	+	+
11:19	2-heptenal	+	+
11:21	benzaldehyde	+	+
11:54	methylfurfural	+	+
11:59	methyl furoate	–	++
12:25	methylheptenon	+	+
12:59	octanal	+	+
13:41	methylheptanol	++	++
13:58	ethlyhexanol	+	+
14:19	phenethylaldehyde	+	+
14:53	2-octenal or nonenal	+	+
16:27	nonanal	+	+
16:53	phenethylalcohol	+	+
17:46	dimethoxybenzene	–	+
18:12	nonenal	+	+
18:46	methyl phenylacetate	–	++
19:40	decenol	+	+
21:58	unknown (192)	+	(–)
22:30	gamma–lacton	++	++
23:11	gamma–lacton	+	+
24:03	unknown	+	+
24:07	allylmethoxyphenol	+	+
24:38	octyl hydroxypropanoate	+	+
25:18	trimethoxybenzene	–	+
26:06	Me-chlorohydroxybenzoate	+	+
26:38	dimethylundecanon	+	+
28:00	unknown (200)	+	(–)
29:03	unknown (200)	+	(–)
29:24	unknown	–	+
29:35	C12 fatty acid	++	++
31:56	furan type	(–)	++
32:16	C14 fatty acid	+	+
33:17	15:1 fatty acid	+	+
33:27	15:0 fatty acid	+	+
34:21	16:1 fatty acid	+	+
34:43	16:0 fatty acid	++	++
36:25	18:2 fatty acid	++	++
36:53	18:2 fatty acid	+	+

Table 3 *L/N extracts of affected oak wood (*Coniophora puteana *(Cp),* Donkioporia expansa *(de),* Fistulina hepatica *(Fh) and a mixture of Fh and De (FhDe), comparison of affected wood extracts, after different time periods.*

compound	Cp8	Cp17	De8	De20	Fh10	Fh17	FhDe
methyl furoate		+	+		+		
dimethoxybenzate		+	+			+	
methyl phenylacetate		+	+			+	
furan-type (190)		+	+			+	
unknowns (homologs)	+	+					
unknown (200)	+	+					
terpene-type (220)						+	+	+
unknown (166)							+
unknown (168)							+

cal sugar decomposition products were detected, such as methylfuroate and two unknown furan-like compounds. In addition, decomposition products which probably originate from the lignin part of the matrix were found: methyl phenylacetate and dimethoxybenzene (see Figs 6 and 7 and Table 2).

In the Cp samples three typical compounds occurred which were not detected in either the sound heartwood or the De-affected heartwood samples. The structure of the compounds with molecular weights of m/z 192 and 200 remains unknown.

After 10–20 weeks of fungal attack the major typical compounds remained the same for De and Cp. After ten and 17 weeks Fh induced the formation of a sesquiterpene-like compound (m/z 220). The mixed fungal growth of *Donkioporia expansa* and *Fistulina hepatica* (DeFh) resulted in the remarkable formation of two unknown compounds with m/z 166 and 168, besides the compounds found in wood affected only by either De and Fh (Table 3).

The volatile compounds from all Scots pine samples appeared to be highly complex and very numerous, not allowing extensive identification of compounds. The relatively low amounts of volatile compounds in European oak may play an important role in the preference of deathwatch beetles for this wood species.

Wood samples

Py-DCI/MS was carried out on finely powdered wood samples. PC-DA of the Py-MS profiles revealed the major trends and differences in the data. The fungus *Fistulina hepatica* shows a large effect on the wood composition after ten weeks time, compared to the composition of sound wood samples (Fig 8, code Ek, European oak heartwood and Es, European oak sapwood). Stretching this period to 17 weeks does not have a further effect as shown by the coinciding positions in the plot (Fig 8, codes Efh10 and Efh17). According to the positions in the plot, De and Cp show the same trend in the degree of decomposition going from sound wood to eight weeks and 17–20 weeks of growth. FhDe, the fungal mixture, has a position next to the De and Fh profiles. Interpretation of the major axis of chemical difference revealed higher intensities of oligohexoses, oligopentoses and acetylated oligohexose compounds in the affected samples. These oligomers (that originate from the cellulose/

samples. Volatile compounds as well as non-volatile compounds were monitored. Chromatograms and spectra of sound and affected wood have been compared.

Samples of affected and sound oak wood were analysed with gas chromatography-mass spectrometry (GC-MS), pyrolysis-direct chemical ionization mass spectrometry (Py-DCI-MS), matrix-assisted laser desorption mass spectrometry (MALDI/MS) and nuclear magnetic resonance spectrometry (NMR). Only Py-DCI/MS could be performed directly on the wood samples. The other analytical techniques were carried out on wood sample extracts. The PC-DA technique was applied in the evaluation of Py-MS and NMR data.

Volatile compounds

GC-MS analysis of affected timber heartwood samples revealed that, after eight weeks, De and Cp, especially De, induced important changes in the pattern of volatile compounds. In comparison with sound heartwood, typi-

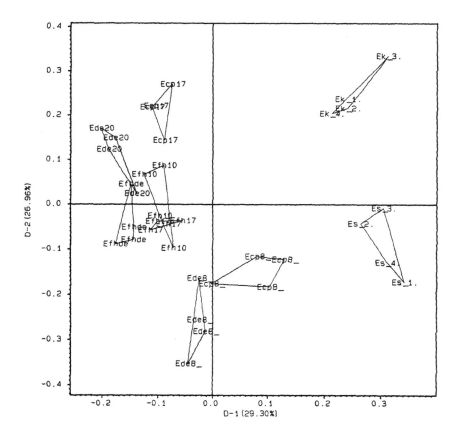

Figure 8 Discriminant plot of Py-DCI/MS spectra (averaged) of sound oak (heartwood = Ek, sapwood = Es) and affected oak (codes cp, de, fh and fhde).

hemicellulose complex) are possibly liberated from the lignin network and therefore better visible in these spectra.

Extracts of wood samples

Water extracts and organic extracts were analysed by NMR. Data analysis of the sound European oak (heartwood, code Ek), and affected samples Cp-, De, Fh and FhDe revealed that the decayed samples contained higher intensities of oligosaccharides. This is in full agreement with the results of Py-MS. Additional indications were obtained from the spectra, that possibly deoxysugars were a product of fungal activity or a decomposition product of the cellulose/hemicellulose complex.

The organic extract highlighted another chemical aspect due to fungal decay. Differences found in the spectra may indicate a diminishing quantity of sitosterol due to fungal attack.

MALDI/MS results did not provide clear spectral differences in the spectra of the water extracts. Possibly too high molecular weight compounds were present in the water extract.

Conclusions

The chemical composition of European oak with fungal decay can be clearly distinguished from sound samples of the same origin. European oak contains a relatively very low amount of volatile compounds, compared to Scots pine.

Donkioporia expansa especially produces a number of unique compounds in relatively high amounts. Furan-

type compounds probably originate from the polysaccharide part of the matrix, whereas phenylacetate and dimethoxybenzene can be metabolites from the lignin part. The same metabolites are produced over an extended time period. Remarkably, a mixture of *D. expansa* and *Fistulina hepatica* (DeFh) produces some unique compounds which were not detected for one or other of the individual fungi.

F. hepatica growth slows down after a short period of time. Possibly not only the ratio between lignin and polysaccharides changes, but also the network architecture of the cellulose/hemicellulose complex.

DIFFERENTIATION OF HEARTWOOD AND SAPWOOD FROM EUROPEAN AND AMERICAN OAK

Introduction

One of the first aspects to be investigated was the differentiation of European and America oak. Differences in chemical composition are of interest because in the literature it is mentioned that American oak is more susceptible to attack by deathwatch beetle than European oak.

Chemical analysis of volatile compounds

GC-MS analysis revealed a remarkable difference in chemical composition between the investigated samples from American and European oak. The compound 3-methyl-4-hydroxyoctanoic acid, gamma lacton could only be detected in the European oak samples. Has this compound an inhibiting effect on the occurrence of deathwatch beetle? American oak is known for its higher susceptibility to decay. The gamma lacton compound is not reduced in the experiments with new European oak by short-term decay (16–20 weeks in the laboratory, see above). The slow evaporation of this low volatile compound may have an influence in the changes measured in aged samples from buildings (see above). This would need further research for confirmation.

CONCLUSIONS

1. Chemical analysis of wood can be used to distinguish early fungal decay in oak invisible to the naked eye.
2. The high intensity of oligomers at insignificant fungal decay (mass loss <3%) of the oak samples from the laboratory indicates a change in the network architecture of the lignin/cellulose/hemicellulose cell wall complex, and may result in an easier digestible food source for deathwatch beetle larvae.
3. Pyrolysis-mass spectrometry of oak from buildings indicates a change in composition with age, but further research and a larger number of wood samples is needed to confirm this.
4. Pyrolysis-mass spectrometry can be used in cases of doubt on the extent of decay in important wood structures, as a fast, reliable and reasonably cheap, low impact assessment method. Remedial treatments may be adjusted according to the results.

BIBLIOGRAPHY

EN 113, 1996: *Wood Preservatives - Test Method for Determining the Protective Effectiveness against Wood-Destroying Basidiomycetes - Determination of Toxic Values.*

Esser P M, 1995 TNO report 95-CHT-R1302, *Woodcare; Understanding the Relationships between Deathwatch Beetle, Wood Decay Fungi and Timber Aging in European Historical Buildings in order to Develop Alternatives to Current Harmful and Ineffective Treatments.* Contract EV5V-CT94-0517.

Fisher R C, 1940 Studies of the biology of the Death watch beetle, *Xestobium rufovillosum* De G. Part III. Fungal decay in timber in relation to the occurrence and rate of development of the insect, in *Applied Biology* **27**, 525–557.

Fisher R C, 1941 Studies of the biology of the Death watch beetle, *Xestobium rufovillosum* De G. Part IV. The effect of type and extent of fungal decay in timber upon the rate of development of the insect, in *Applied Biology* **28**, 244–259.

AUTHOR BIOGRAPHIES

Petra Esser is leader of the Wood and the Environment working group in the Centre for Timber Research at TNO Building and Construction Research in The Netherlands. She has an MSc degree in Environmental Biology from the University in Leyden and over 10 years of experience in biological and environmental research. Research topics range from degradation processes of wood, effectiveness of anti-sapstain treatments and impregnation of wood, measuring of emissions from preservative treated wood, to performance of complete life cycle assessments of wooden products. Dr Esser was the TNO coordinator of the Woodcare project and prepared laboratory and field wood samples for chemical analysis.

Albert Tas is Product Manager Structure Elucidation & Pattern Recognition Research of the Division of Analytical Sciences at TNO Nutrition and Food Research Institute. He has over 20 years experience in organic chemistry and spectrum interpretation and circa 10 years in multivariate data analysis (MVA) of complex spectral data. Multivariate data analysis of complex spectral data has been the subject of his thesis "Mass Spectrometric Fingerprinting: Soft Ionization and Pattern Recognition", finished in 1991. Current research topics are ripening processes of food products, the determination of authenticity and quality of food products, and impurity profiling of fine chemicals. His role in the Woodcare project was the coordination of chemical analyses and interpretation of spectral data from GG-MS, Py-DCI/MS and NMR supported by MVA.

The effect of fungi on the growth of deathwatch beetle larvae and their ability to attack oak

BRIAN V RIDOUT AND ELIZABETH A RIDOUT
Ridout Associates, 147A Worcester Road, Hagley, West Midlands, DY9 0NW, UK;
Tel: +44 (0)1562 885135; Fax: +44 (0)1562 885312; email: ridout-associates@lineone.net

ABSTRACT

It had been shown previously that wood-decaying fungi accelerated larval growth of the deathwatch beetle in oak. This was thought to be due to the concentration of available nitrogen during decay. This study investigated this theory using protein/amino acid assay. No evidence was found that increased beetle activity was correlated with an increase in available nitrogen, or that the nitrogen content of infested heartwood was significantly greater than that of sound heartwood.

Key words

Deathwatch beetle, larval growth, fungus, nitrogen

INTRODUCTION

In 1924 Maxwell-Lefroy stated that deathwatch beetle larvae could burrow into the sapwood or 'solid heartwood' of oak, and would tunnel from one timber into the next, regardless of the relative positioning of grain within the two timbers. Kimmins, who had assisted Maxwell-Lefroy until the latter's death in 1925, modified these observations by stating that the beetles would not attack freshly seasoned oak. He suggested that this might be because the moisture content was too high, or because of the absence of 'fungoid growths' (Kimmins 1933, 133). The latter suggestion followed from observations made by Munro (1931) who was Maxwell-Lefroy's first assistant, and had found deathwatch beetle larvae in stag-headed oak trees in Richmond Park. These were large trees of considerable age (Fig 1) and the beetles were found in the dead wood. Munro observed that the stag-headedness was reported to be frequently associated with fungus (Braid 1924) and that the beetle might therefore require fungus in oak wood before it could be attacked. The observation was consistent with earlier records of the larvae from decayed hardwoods (eg Westwood 1839-40). Fisher (1940) studied building timbers where, as reported by Maxwell-Lefroy, there were beetles but apparently no fungal decay. He cut sections for microscopic examination, and found that fungus was always present. He also showed that the type of fungus was unimportant, and that the length of the larval growth

Figure 1 A stag-headed oak tree in Richmond Park, London, photographed by Munro in 1913 (D Dickinson, Imperial College London).

Figure 2 The light-coloured oak sapwood is attacked by deathwatch beetle, while the darker heartwood is undamaged. (John Fletcher). See also Colour Plate 11.

period was proportional to the weight loss in timber caused by the fungus. Sapwood and heartwood of modern oak were not attacked unless decay was present, but attack could progress slowly in undecayed old oak heartwood. Fisher (1941) provided more experimental data to confirm his earlier findings, and reiterated that the larvae would not attack sound oak sapwood. The age of the sapwood is not given, but where old oak is used in previous experiments (Fisher 1940) this fact is stated. It seems likely therefore that the oak sapwood used was modern. Observations certainly suggest that old oak sapwood can frequently be attacked when oak heartwood remains untouched (Fig 2), but the wood has never

been examined for fungus with a microscope in these situations. Modern oak may have too many volatiles to be acceptable to these beetles, and this would be in agreement with Hickin (1975), who believed that oak must mature for about 60 years before an infestation could commence.

The necessity for fungus in the heartwood of building timbers before they can be successfully colonised by deathwatch beetle seems understandable if the insects' natural food source is never free from fungus. The role that the fungus takes in turning heartwood into an acceptable beetle food source is the subject of this investigation.

THE EFFECTS OF THE FUNGUS

Fisher (1940) found that the effects of decay on larval growth were most pronounced at the lower end of the decay scale, and that the length of the insect life cycle in oak was not appreciably affected when the extent of the decay exceeded 45% (Fig 3). These results, backed by chemical analysis data from his colleagues Campbell and Bryant (1940) which showed that the larval diet did not contain recognisable amounts of fungal components, led Fisher to conclude that the primary effect of the fungus was to reduce the mechanical strength of the timber. This would enhance the conservation of energy during the metabolism of the insect by making the wood easier to attack, and thus, presumably, enhance the growth rate of the larvae. The effect would be particularly beneficial to newly-hatched larvae. Fisher did however remark in the same paper that the fungus might cause other changes in the timber which could not be detected by the methods of analysis then available. This led to the suggestion that the fungus might also provide a rich nitrogen source, and he quoted Becker (1938) who showed that the addition of proteins to pine sapwood considerably enhanced the growth rate of house longhorn beetle (*Hylotrupes bajulus*). Fisher's suggestions and Becker's researches persuaded Campbell and Bryant to modify their conclusions (Campbell 1941). They now believed that the fungus removed major components from the timber at a greater rate than nitrogen, so that net nitrogen content increased

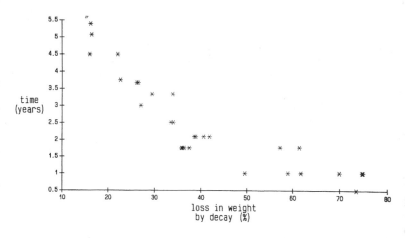

Figure 3 Length of insect life cycle and extent of decay.

as the timber decayed and the nitrogen became more accessible to the larvae.

Several authors have shown that the nitrogen content of timber drops from the cambium through the sapwood, and that larger larvae may be grown from the outer sapwood (Becker 1977). A correlation between these two factors has apparently been obtained by experimenting with the extraction and addition of nutrients (ibid). Nevertheless the quantity of available nitrogen in wood is usually less than 0.1%, according to Fengel and Wegener (1989), but only small amounts are apparently essential to the larvae: Bletchley (1966) showed that the minimum nitrogen content for the growth of the furniture beetle (*Anobium punctatum*) larvae was 0.03%.

ASSESSMENT OF NITROGEN CONTENTS IN ACTIVE INFESTATIONS

At the time when Bletchley and Becker undertook their work, nitrogen in timber was assessed by the Kjeldahl method, and this was seen by Bletchley (1969) to be a disadvantage. In this method the wood is digested in acid and the total nitrogen content is calculated, but some of the nitrogen may be in a form unavailable to the larvae. Nowadays assay techniques which measure protein and amino acid contents can provide an estimation of available nitrogen in forms which the larvae can assimilate. The role which the fungus plays in larval development can now be assessed in a little more detail.

Methods

The roofs of two buildings were chosen for an infestation assessment, and tissue paper was fastened with water-soluble glue over groups of beetle emergence holes. Current activity was determined by examining the papers for flight holes at the end of the emergence season. Samples of timber were removed with a 6 mm (¼ in) auger bit from localized areas where current infestation was demonstrated. These samples were removed from the surface to a depth of 50 mm (2 in), ground and subsampled to provide 150 mg of milled dust. The dust was mixed with 1 ml of water in an Eppendorf tube and rotated at 4^0 C overnight. The extract was centrifuged at 13000 rpm for five minutes and then analysed by the Comassie blue dye binding method using Comassie® Plus Protein Assay Reagent from Pierce. The standard protocol was used, adding 1.5 ml reagent to 50 ml sample. Three mg/ml PVP was added to the reagent to bind any phenolic compounds present. Absorption was read at 595 nm and the protein concentration (mg/ml) was determined from a standard curve prepared with bovine serum albumin. Three experiments were undertaken.

Experiment 1. The Tide Mill, Beaulieu, Hampshire, seventeenth-century timbers

Replicate samples were removed from one location (150 mm²), where little activity was found (one hole) during the 1997 emergence season and compared with samples from three similarly sized locations where multiple holes

were found. In each case the sample was taken from within 25 mm (1 in) of a fresh emergence hole.

Experiment 2. The Vyne, Hampshire (1640)

Zero emergence one year may not necessarily mean zero emergence the following year. Samples were therefore taken in a similar fashion to experiment 1 from The Vyne, where emergence data for two consecutive years were available. Sample replication was not undertaken because damage to the timber had to be minimised.

Experiment 3. Sound heartwood

The previous experiments compared samples from sites where there were different levels of beetle activity, but undamaged timber was not sampled. Fisher (1940), however, found that larval development was possible in sound historical heartwood, although at a very slow rate. It is possible that complex molecules might break down over the centuries making more nitrogen available to larvae in ancient timbers. Fifty samples from dendrochronologically dated heartwood with a twelfth- to twentieth-century span were analysed for available nitrogen in experiment 3. Each sample was approximately a 20 mm cube, and was taken from midway across the heartwood of a transversely cut section of timber.

Results

Available nitrogen contents from the Tide Mill samples are listed in Table 1, and those from The Vyne experiment in Table 2. Neither gives any indication that increased beetle activity is correlated with an increase in nitrogen content.

Data from the undamaged heartwood are presented in Table 3 where the median age is taken to be the approximate age of the sample because of the sampling position. There are no suggestions from this data that available nitrogen increases or decreases with the age of the sample. As a result high values and low values are mixed together in every century. The mean nitrogen content from all of the samples where beetle damage was present (Tables 1 and 2 [0.13±0.071 (n=15)]) was compared with the mean from the undamaged heartwood samples (Table 3 [0.18±0.095 (n=46)]). There was no significant difference (P<0.20 with 69 d.f).

There is no evidence from these data that the concentration of nitrogen by fungus or an increased nitrogen availability with time makes oak heartwood susceptible to attack.

Table 1 A comparison of available nitrogen in replicated samples from timber with different levels of beetle activity at the Tide Mill, Beaulieu, Hampshire.

Number of beetles emerged (16.04.97–14.07.97)	% available nitrogen
1	0.15±0.081 (n=3)
3	0.13±0.051 (n=3)
5	0.37±0.249 (n=3)
6	<0.072±0.04 (n=5)

Table 2 A comparison of available nitrogen in single samples from timber with different levels of beetle activity at The Vyne, Hampshire.

Number of beetles emerged		% available nitrogen
1996	1997	
0	0	0.13
1	0	0.10
1	0	0.10
1	1	0.10
1	1	0.15
1	2	0.10
3	0	<0.09
3	1	0.12
0	4	0.10
1	4	0.10
14	10	<0.09
29	17	0.12

Table 3 A comparison of available nitrogen in mature oak heartwood tabulated according to the median century of growth.

sample number	timber growth period	median age	% nitrogen
1	1174–1283	1228	0.18
4	1230–1300	1265	0.33
5	1220–1340	1280	0.09
6	1168–1271	1219	0.32
22	1220–1397	1308	0.09
23	1260–1320	1290	0.13
2	1300–1350	1325	0.09
3	1301–1381	1341	0.09
24	1307–1377	1342	0.14
25	1318–1406	1362	0.10
46	1331–1430	1380	0.10
8	1384–1427	1405	0.09
10	1435–1530	1482	0.10
26	1375–1450	1412	0.17
27	1395–1460	1427	0.13
28	1348–1528	1438	0.14
29	1387–1585	1486	0.50
30	1435–1539	1487	0.12
47	1385–1524	1454	0.17
48	15th century (d.p.★)		
31	1409–1592	1500	0.09
11	1460–1570	1515	0.12
12	1513–1606	1559	0.14
14	1534–1637	1585	0.34
32	1513–1610	1561	0.18
33	1463–1627	1545	0.25
34	1540–1640	1590	0.18
13	1575–1650	1612	0.14
15	1666–1722	1694	0.13
16	1655–1739	1697	0.13
19	17th century (d.p.★)		
20	17th century (d.p.★)		
35	1660–1723	1691	0.16
49	17th century (d.p.★)		
50	17th century (d.p.★)		
36	1670–1739	1704	0.19
37	1675–1735	1705	0.12
38	1630–1770	1700	0.12
39	1710–1801	1755	0.12
40	1749–1806	1775	0.31
41	1651–1810	1730	0.17
42	1725–1814	1769	0.21
18	1765–1860	1812	0.48
43	1800–1835	1817	0.18
45	1756–1981	1868	0.17
44	1859–1950	1904	0.19

★d.p. dated by provenance
★★ when there is more than one sample with the same median age then they are ordered by the earliest date at the beginning of their age range

CONCLUSIONS

There is no evidence from these data to support the theory that fungi aids deathwatch beetle attack by concentrating available nitrogen. Neither is there evidence that available nitrogen content increases as the timber ages, which might have explained Fisher's observations that the larvae could grow slowly in historical heartwood without the presence of fungus.

The idea that fungi would concentrate nitrogen and thus assist the growth of beetle larvae is an attractive one. Bletchley and Farmer (1959), for example, considered that nitrogen content was one of the most important factors determining suitability for attack by the closely related furniture beetle (*Anobium punctatum*) and Martin (1979) quotes many references to illustrate that fungi concentrate nitrogen. He also quoted Merrill and Cowling (1966), who showed that a larva would have to consume 36.2g of wood infected with *Ganoderma applanatum*, in order to obtain the same nitrogen content as 2.7g of the *G. applanatum* sporophore. The nitrogen content of fungus mycelium is not necessarily high, however, and Cowling and Merrill (1966), working with the wood-rotting fungi *Fomes larcis*, *Fomes fomentatius* and *Coriolus versicolor*, found a nitrogen content in the mycelia of 0.23% to 3.27%. The low end of the range is no greater than the nitrogen content we found in many undecayed sample of oak. It is certainly possible that a significant quantity of fungus mycelium might increase available nitrogen levels if consumed, or in the form of enzymes secreted into the surrounding wood, but we found no indications of this in our samples.

Fungi might also supply other nutrients, and Norris (1972) for example showed that the sterol requirement of the wood-boring ambrosia beetle *Xyleborus ferrugineus* was obtained by consumption of the symbiotic fungus *Fusarum solani*. If, however, there was not enough mycelium present in our samples to affect nitrogen content, then perhaps there is not enough to influence any other nutrients. Martin (1979) suggested that fungus enzymes assimilated from a substrate might increase the digestive capabilities of an insect consuming them. This would require very stable enzymes and a compatible pH range in the insect's digestive tract.

What seems to be necessary is a significant change to the wood caused by a small amount of fungus. The most likely role the fungus plays in aiding a beetle attack is that of modifying the chemistry of oak heartwood. Esser and Tas (this volume) have shown that a visually undetectable amount of fungus can significantly change wood chemistry. The fungus may detoxify repellant or harmful compounds. It may also split complex molecules into units that are easier for larvae to assimilate.

BIBLIOGRAPHY

Becker G, 1938 Zur Erriahrungsphysiologie der Hausbockkäfer-larven (*Hylotrupes bajulus* L.) *Naturziss* **26**, 462–463.

Becker G, 1977 Ecology and physiology of wood-destroying Coleoptera in structural timber, in *Material und Organismen* **12**: 3, 141–160.

Braid T, 1924 Some observations on *Fistulina hepatica* and hollow stagheaded oaks, in *Transactions of the British Mycological Society,* **9**, 210.

Bletchley J D, 1966 Aspects of the habits and nutrition of the Anobiidae with special reference to *Anobium punctatum* de Geer, in *Beihefte zu Material und Organismen* (Internationales Symposium Berlin-Dahelm 1965). Supplement to *Materials and Organisms* **1**, 371–381

Bletchley J D, 1969 Seasonal differences in nitrogen content of Scots Pine (*Pinus sylvestris*) sapwood, and their effects on the development of the larvae of the common furniture beetle (*Anobium punctatum*), in *Journal of the Institute of Wood Science* **22**:4, 43–47.

Bletchley J D and Farmer R H, 1959 Some investigations into the susceptibility of Corsican and Scots pines and European oak to attack by the Common Furniture Beetle (*Anobium punctatum*), in *Journal of the Institute of Wood Science,* **3**, 2–20.

Campbell W G and Bryant SA, 1940 A chemical study of the bearing of decay by *Phellinus cryptarum* Karst. and other fungi on the destruction of wood by the deathwatch beetle, *Xestobium rufovillosum* de Geer, in *The Biochemical Journal,* **34**, 1404–1414.

Campbell W G, 1941 The relationship between nitrogen metabolism and the duration of the larval stage of the deathwatch beetle (*Xestobium rufovillosum* De Geer) reared in wood decayed by fungi, in *The Biochemical Journal,* **35**, 1200–1208.

Cowling E B and Merrill W, 1966 Nitrogen in wood and its role in wood deterioration, in *Canadian Journal of Botany,* **44**, 1539–1587.

Fengel D and Wegener G, 1989 *Wood Chemistry, Ultrastructure, Reactions.* Walter de Gruyter, Berlin, New York

Fisher R C, 1940 Studies of the biology of the deathwatch beetle *Xestobium rufovillosum* De Geer Part III: Fungal decay in timber in relation to the occurrence and rate of development of the insect, in *Annals of Applied Biology,* **27**, 545—557.

Fisher R C, 1941 Studies of the biology of the deathwatch beetle *Xestobium rufovillosum* De Geer: Part IV. The effect of type and extent of fungal decay in timber upon the rate of development of the insect, in *Annals of Applied Biology,* **28**, 244–259.

Hickin N E, 1975 *The Insect Factor of Wood Decay*, third edition, Rentokil Library, London.

Kimmins D E, 1933. Notes on the life history of the deathwatch beetle, *Proceedings of the South London Entomology and Natural History Society,* 133–137.

Martin M M (1979) Biological implications of insect mycophagy, in *Biological Review,* **54**, 1–21.

Maxwell Lefroy H, 1924, The treatment of the deathwatch beetle in timber roofs, in *Journal of the Royal Society of Arts,* **52**, 260–266.

Munro J W, 1931 Insects injurious to timber, in *Journal of the British Wood Preserving Association,* **1**, 51–70.

Westwood J O, 1839–40 *Introduction to the Modern Classification of Insects,* London, Longman.

ACKNOWLEDGEMENTS

The authors wish to thank Professor Monique Simmonds for the use of laboratory facilities at the Jodrell Laboratory, Royal Botanic Gardens, Kew, and Marianne Eyule for undertaking the analysis.

AUTHOR BIOGRAPHIES

Brian Ridout is a Director of Ridout Associates, specialising in complex damp and decay investigations, expert witness work, scientific research and lecturing. He holds degrees from the Universities of Cambridge and London in entomology and mycology, and is a Fellow of the Institute of Wood Science.

Elizabeth Ridout is a Director of Ridout Associates, undertaking decay surveys, collection and analysis of environmental data and the design and installation of remote sensing systems. She holds degrees from the University of Cambridge in applied biology and the analysis of biological data.

Part III

Investigation

The acoustic detection of deathwatch beetle (*Xestobium rufovillosum*) larvae in oak structural timbers

Piet Van Staalduinen, Petra Esser * and Jan de Jong
TNO Building and Construction Research, PO Box 49, 2600 AA Delft, The Netherlands;
Tel: + 31 15 284 2000; Fax: + 31 15 284 3990.

ABSTRACT

A method was developed for acoustic non-destructive detection of active infestation by deathwatch beetle larvae in wooden building constructions.

KEY WORDS

Deathwatch beetle, detection methods, monitoring, treatment

INTRODUCTION

Assessing the activity of deathwatch beetle larvae in the structural timbers in buildings is very difficult. Flight holes made by adult beetles are the only visual marks in the wood, and from the flight holes alone it is not possible to establish if infestations are current. Newly-emerged beetles may give an indication, but only during the flight season (March to June). The direct detection of larvae, which cause the actual damage by eating the wood from the inside, has not been reported.

OBJECTIVES

The aim was to investigate the feasibility of using monitoring methods for detecting larval activity in structures. These methods would assist in planning preventive measures or in assessing the effectiveness of treatment. The research comprised the following steps:

- review of the literature for previous experience with the above mentioned methods and possible alternative techniques. Review of feasibility for application in practice.
- laboratory experiments using larvae for one or more selected techniques
- evaluation of detection thresholds, stability etc, of techniques
- full-scale experiment using one or more of the techniques. The full-scale experiment was to be applied to a structure known to contain larvae.

LITERATURE REVIEW

Several alternatives have already been investigated and evaluated for detecting larval activity in wood, including radar, sonic, ultrasonic and radiation techniques.

Results from a test using impulse radar on wooden beams indicated that only high moisture contents would be detected. Degradation of the inside of the wood due to larval activity could not be detected. Impulse radar could serve only as an indicator for wet spots in wood structures, not for detecting the presence or activity of larvae, nor for detecting deterioration of the wood.

Experience from previous research projects at TNO has shown that the presence of larvae or beetles in separate beams of wood can be assessed by means of gamma radiation. This method is very accurate, but it cannot be applied to complete structures without considerable precautions to control the radiation. A radiation technique was used in this project for assessing larvae or beetle presence in small pieces of wood.

A literature survey was conducted to gain an insight into the possibilities of sonic and ultrasonic methods for detecting larval activity. It was concluded that sonic/ultrasonic detection in the frequency range of 10–40 kHz was feasible, at least for the detection of termite activity (Fujii *et al* 1990a and b, Noguchi *et al* 1991, Fujii *et al* 1992). The technique has also been successfully applied for detecting the activity of the larvae of the house longhorn beetle (Plinke 1991).

FEASIBILITY ASSESSMENT OF SONIC DETECTION

Pilot test set-up

A test set-up was put together consisting of two B&K piezoelectric acceleration transducers (B&K 4384, sensitivity approximately 10 pC/g, measuring frequency range 4–15 kHz), conditioning equipment (B&K 2626), additional band pass filtering (5–30 kHz, Krohn-Hite filters), additional amplification and a measurement computer containing a fast A/D converter board (Nicolet BE 490) on a 66 MHz 486 PC. The detection threshold of this system is approximately 5×10^{-4} g. The data-acquisition software enables monitoring at given trigger levels, and above these the relevant signal is stored on a hard disk.

Initial laboratory monitoring

Based on the findings of the literature search, a test set-up was made in the laboratory, consisting of two pieces of wood, known to have contained larvae of the deathwatch

* Author for correspondence

beetle. Both specimens were connected to a sensitive piezoelectric acceleration transducer detector. The first tests were conducted in a normal laboratory climate (17–19°C, RH 40%). To improve conditions for the possibly active larvae, the specimens were placed in a climatic chamber conditioned to 15°C and 80% RH. However, the test results were not fully decisive.

For this reason, additional monitoring was done on small samples of wood, known to contain larvae of the house longhorn beetle. This monitoring confirmed that the larvae's eating activities induce very small stress waves in the wood, well above the detection threshold of the system. The signals showed a characteristic transient pattern in the time domain and a rather invariable spectrum in the frequency range. Samples containing the larvae of the common furniture beetle (*Anobium punctatum*) were also investigated and it was noted that the signals of the common furniture beetle in general contained higher frequencies (20 kHz range) than those of the house longhorn beetle *Hylotrupes bajalus* (10 kHz range).

Samples of oak containing larvae of the deathwatch beetle were supplied by Birkbeck College, London in the summer of 1995 for testing the initial set-up of the monitoring equipment. Monitoring of these samples gave negative results because the larvae were not active. The activity of the deathwatch beetle was successfully measured using a second series of samples, supplied by Birkbeck in February 1996. This meant that the larvae of three main types of insects affecting wood (house longhorn beetle, deathwatch beetle, common furniture beetle) could be detected under laboratory conditions and could be identified on the basis of measured signals and spectra.

Initial full-scale measurements

The pilot set-up of the monitoring equipment was tested on the basis of one-day monitoring exercises in the church of Naaldwijk which was infested with deathwatch beetle. From this monitoring project, it was concluded that the sensitivity of the equipment to irrelevant disturbances and to disturbances of the electromagnetic field should be diminished.

DEVELOPMENT OF TEST METHOD AND EQUIPMENT

Required equipment

An evaluation on the basis of market availability of alternative components was carried out. The outcome was that B&K 4394/4397 transducers were tested. These transducers have a bandwidth covering 1 to 25000 Hz, which was sufficient for the monitoring purpose. Signals could be amplified using B&K WB 1328 amplifiers. After evaluation of this equipment, it appeared that additional amplifying would be necessary, up to 100 or 1000 times.

Isotron Acceleration Transducers, model 7254-A500, made by Endevco, were also considered in combination with a Signal Conditioner, type 102, from Endevco. This set had the advantage of a wider frequency range, up to 30 kHz, but the signal conditioner needed to be modified to increase the overall amplification. After a final comparison with the specifications of the B&K equipment, this transducer/amplifier set was ordered. Delivery was seriously delayed, due to the procedures required to get the CE conformity mark.

Laboratory measurements

The monitoring equipment was tested on old oak samples known to contain active larvae. Tests were carried out to check the sensitivity of the system for unwanted disturbances. The equipment's sensitivity for low frequency vibrations had decreased, compared to the transducers and amplifiers previously in use, and its sensitivity in the high frequency range (20–30 kHz) had increased.

A number of buildings in The Netherlands were visited during 1996–97 to carry out in situ monitoring work.

- Oosterkerk (church) in Middelburg; active infestation by deathwatch beetle
- Churches in Berlikum, Franeker and Sloten; accessible wooden structure, active infestation by deathwatch beetle
- Church in Goes; active infestation by deathwatch beetle.
- Trippenhuis, Amsterdam, uncertain whether or not active infestation was present, wooden structural parts have been infested in the past
- Gevangenenpoort, The Hague; active infestation by common furniture beetle and deathwatch beetle.

Finally, controlled calibration measurements were made on 20 and 21 January 1997 at the Royal Botanic Gardens, Kew, UK, on samples of wood known to contain the larvae of the deathwatch beetle. In addition, the equipment was also tested in the UK by making measurements in two heavily infested structures, Winchester Cathedral and Bishopstone Church. These measurements took place on 22 and 23 January 1997.

The aims of the measurements performed at Kew were to:

- validate the monitoring system
- fine-tune the system parameters
- evaluate alternatives for the connection between specimen and transducers.

The measurements performed in Bishopstone Church and Winchester Cathedral were to validate the system in practice.

The measurements at Kew were very successful. Wooden samples containing larvae, checked by taking X-ray photographs some weeks earlier, could be positively identified. Activity of larvae as young as approximately eight months could be demonstrated. The activity rate of the larvae was established at various temperatures, both day and night. A strong dependency on temperature was apparent. Most noteworthy, even at temperatures as low as 6°C some activity could be shown (Steenhuis *et al*

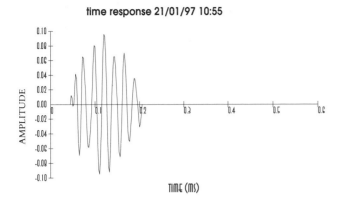

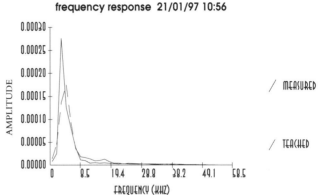

Figure 1 Results of measurement KEW8112.M1.

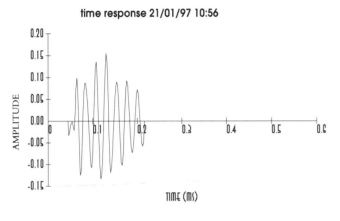

Figure 2 Results of measurement KEW8113.M1.

1997). Figures 1 and 2 show examples of measured vibration signals and spectra, compared to previously entered reference spectra on the basis of the laboratory experiments.

The following conclusions can be drawn from the monitoring sessions at Kew:

- larvae of death watch beetles can be detected successfully by ultrasonic measurements in a frequency range from 1000 to 40000 Hz
- in specimens ranging from one to 13 years old, activity could be detected. The resulting spectra are not dependent on the age of the specimen
- test samples without deathwatch beetle larvae did not trigger a signal, which shows the reliability of the system.

SOFTWARE

Data acquisition

The software required for data acquisition is commercially available and runs on a PC. Data acquisition software consists of Team490 software version 3.20A. The following functions can be performed with the help of this software:

- waiting for next trigger
- triggering of the measurement
- data storage of the measurement.

Analysis software

The starting point for the development of the analysis software was the initial laboratory experiment. In the laboratory the signals from two transducers were measured simultaneously. The signals were only recorded once the signal amplitude exceeded a certain pre-set level (a trigger level). Since external (unwanted) influences are likely to affect the signals of both transducers, the software processed only those data where one transducer measured activity, and the other transducer measured no activity. This was to eliminate false readings.

The analysis software we developed is self-teaching. It compares new measurements to previous measurements by determining a correlation coefficient between the shape of the vibration spectra. Then the program decides between true (good agreement between new and old data) and false readings (bad agreement between new and old data). The old data is represented by a weighted average of the accepted previous measurements. The starting values are represented by measurements from the laboratory experiments.

correlation =

$$\frac{(\Sigma\ (x-\sigma_x)(y-\sigma_y))^2}{\Sigma(x-\sigma_x)(x-\sigma_x).\Sigma(y-\sigma_y)(y-\sigma_y)}$$

where:

x   is the spectrum of the measured signal in a discrete point;
σ_x is the average of the spectrum of the measured signal;
y   is the spectrum of the reference signal in a discrete point;
σ_y is the average of the spectrum of the reference signal;
Σ is the summation over all discrete points in the spectrum;

The correlation equals to one in case of a perfect fit of the measured spectrum and the reference spectrum.

Figure 3 Correlation equation.

Additional software for performing statistics on the measured data has been prepared. The purpose of this is to obtain knowledge about the activity of the insects over a certain time span (eg a day).

The analysis software performs some important functions to check the quality of the measured data. The following functions are carried out:

- reading the measured signals stored by Team490, with two response functions, one for each channel
- filtering the time response functions, vibrations well below the trigger level are given a null value
- detecting whether the two channels trigger at the same moment. If so, the program is terminated without further evaluation as the trigger shows that an external cause other than the activity of larvae is involved
- transforming the time response function of the triggering signal into a spectrum with the help of Fast Fourier Transform action (FFT).
- comparing the spectrum to a 'reference spectrum', considered to be ideal for the activity of larvae of a certain insect species. This comparison is made by means of the correlation calculation in Figure 3.

Control and monitoring software

The software developed to control the measurements and to present the analysis results has three options. It can be used to control and follow a monitoring session, in which case the monitoring hardware is connected to the PC. Secondly, it can be used to re-evaluate measured data from a previous monitoring session. Thirdly, it can be used to browse through an existing set of monitoring data. The software operates in an Windows environment.

To follow a monitoring session
The software to control and to present the monitoring session performs the following functions.

- it controls the triggering and storage process of the Team490 software
- it starts the evaluation process of the analysis software: based on the outcome of this program, the software overwrites the last measurement with a new one or stores the last one

- the results of the monitoring session can be monitored visually.

To re-evaluate a monitoring session
The software can re-evaluate results produced by Team490 software with other settings. In that case, it performs the following functions:

- it starts the evaluation process of the analysis software. Based on the outcome of this program, the software can skip the last measurement or store it.
- the results of the monitoring session can be monitored visually.

To browse through a existing set of monitoring data
The software's main functionality is to browse through an existing set of data. The following information is presented to the user:

- the time response function of the measured signal after filtering with the analysis software
- the spectrum of the measured signal in comparison with the reference spectrum
- trigger statistics: an overview of the activity of the larvae throughout the whole monitoring period
- a report presenting the trigger statistics alphanumerically.

The measurements can be performed by people without background knowledge of the equipment. Figures 4 and 5 show some impressions of the user interface of the control software.

CONCLUSIONS

Sonic/ultrasonic detection can be used to assess the rate of activity of deathwatch beetle larvae and thus develop more tailored and targeted treatment schedules. Early detection of eight months old larvae is possible. Room temperature should be 10°C or higher.

Sonic/ultrasonic detection can be used after treatments to assess their efficacy.

BIBLIOGRAPHY

Fujii Y, Imamura Y, Shibata E and Noguchi M, 1992 Feasibility of AE monitoring for the detection of the activities of wood-destroying insects, *23rd Annual Meeting, International Research Group on Wood Preservation*, WG II, Harrogate, UK, May 1992.

Fujii Y, Noguchi M, Imamura Y and Tokoro M, 1990a Using acoustic emission monitoring to detect termite activity in wood, in *Forest Products Journal*, **40**:1, 34–36.

Fujii Y, Noguchi M, Imamura Y and Tokoro M, 1990b Detection of termite attack in wood using acoustic emissions, in *21st Annual Meeting, International Research on Wood Preservation*, WG II, Rotorua, New Zealand, May 1990.

Lewis V R, Lemaster R L, Beall F C and Wood D L, 1991 Using AE monitoring for detecting economically important species of termites in California, in *22nd Annual Meeting, International Research Group on Wood Preservation*, WG II, Kyoto, Japan, May 1991.

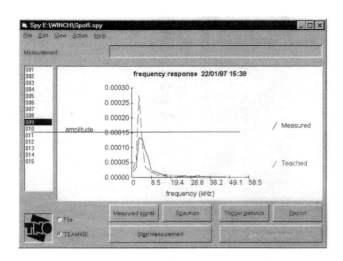

Figure 4 Layout of the user interface of the spy software (impression 1).

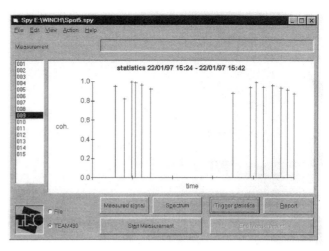

Figure 5 Layout of the user interface of the spy software (impression 2).

Noguchi M, Fujii Y, Owada M, Imamura Y, Tokoro M and Tooya R, 1991 AE monitoring to detect termite attack on wood of commercial dimension and posts, in *Forest Products Journal*, **41**:9, 32–36.

Plinke B, 1991 Akustische Erkennung von Insektenbefall in Fachwerk, *Holz als Roh- und Werkstoff*, 49, 404.

Steenhuis C M, 1997 *Monitoring of death watch beetles carried out at Kew Royal Botanic Gardens, Bishopstone Church and Winchester Cathedral*, TNO report 97-CON-R0791, 20 August 1997.

AUTHOR BIOGRAPHIES

Piet van Staalduinen is head of the department of Structural Dynamics in TNO Building and Construction Research, The Netherlands. He graduated in 1986 from Delft University of Technology, Department of Civil Engineering. He has been working in research and advisory projects concerning the dynamic behaviour of buildings, safety and reliability assessment of buildings and structures and developments of non-destructive testing methods to assess properties of structures. In the Woodcare project he was responsible for the development of the detection method of larvae and for additional laboratory and *in situ* tests.

Petra Esser is leader of the Wood and the Environment working group in the Centre for Timber Research at TNO Building and Construction Research in The Netherlands. She has an MSc degree in Environmental Biology from the University in Leyden and over 10 years of experience in biological and environmental research. Research topics range from degradation processes of wood, effectiveness of anti-sapstain treatments and impregnation of wood, measuring of emissions from preservative treated wood, to performance of complete life cycle assessments of wooden products. Dr Esser was the TNO coordinator of the Woodcare project and prepared laboratory and field wood samples for chemical analysis.

Jan de Jong is leader of the working group Wood in Buildings in the Centre for Timber Research of TNO Building and Construction Research, The Netherlands. His educational background is in building technology. He has 15–20 years of experience of working at TNO on damage assessment in wooden structures and wooden building interiors. He also advises on repairs and treatment of wood and is involved in testing the durability of products and product development. Projects are performed for builders, industrial clients, government (national and the EU) and building co-operatives. His role in the Woodcare project was monitoring conditions in buildings, detection of deathwatch beetle larvae in laboratory and field conditions (with Piet van Staalduinen) and testing heat treatments.

Non-destructive location and assessment of structural timber

ROBERT DEMAUS

Demaus Building Diagnostics Ltd, Stagbatch Farm, Leominster, HR6 9DA, UK;
Tel: +44 (0)1568 615662; Fax: +44 (0)1568 615659; www.demaus.co.uk

Abstract

In the repair and conservation of historic buildings, ever greater emphasis is put on the need to minimise disturbance to, and loss of, original or early material. While this approach is undoubtedly commendable, it can often create conflicts with fundamental obligations to maintain or reinstate structural integrity, and to minimise the processes of degradation. When other relevant factors such as economic constraints and indemnity against liability are included, the need for cost-effective and accurate non-destructive diagnosis of perceived and actual faults becomes paramount. This paper discusses three non-destructive techniques that have proved the most effective for *in situ* location and assessment of timber, and explains their applications.

Key words

Timber decay, diagnostics, non-destructive testing, thermography, micro-drilling, ultrasound

INTRODUCTION

The demands put on those charged with the repair and conservation of our built heritage continue to expand in many and conflicting directions. On the one hand, there is a growing perception that all repair or intervention should be kept to an absolute minimum. On the other, increased expectations of performance, safety and longevity (English Heritage itself frequently requests 40–50 year repair cycles on grant-aided work), coupled with cost restraints and concerns over professional indemnity, may frequently be at odds with this approach. There is therefore an increasing demand for methods of assessment that not only provide more comprehensive and more accurate data on factors affecting structural integrity, stability and degradation, but do so with no, or minimal, damage, or disturbance to any part of the fabric: they must also be cost-effective. Over the last few years several techniques have been developed that can achieve these demands and bridge the gap between these conflicting requirements.

The time when it was acceptable to pass or fail a piece of structural timber merely by visual inspection, or a prod with an old screwdriver, has (or should have) passed. Of course, there are timbers that are so obviously totally decayed, or in such overtly good condition (always assuming that they are visible in the first place), that no further assessment is necessary, but most timbers fall between these two extremes. All too often, timbers that could be retained, perhaps with some minor repair, are rejected or subjected to major repairs because inadequate assessment has suggested that they are no longer capable of fulfilling their function. Conversely, timbers that should be repaired, or replaced, are missed because the degradation occurs away from the visible or accessible faces.

The dimensions of the majority of structural timbers in historic buildings exceed structural requirements, and so contain a considerable margin of safety: a certain level of degradation can therefore be allowed, but it is essential that its extent and position within the cross-section of the timber is known. For example, the juvenile wood toward the centre of a boxed heart section is particularly vulnerable to fungal and deathwatch beetle attack, but this may not be of structural significance provided it remains contained within the central section. However, if (as often happens) the heart deviates from the central axis of the element, such degradation could become critical, but remain undetected by visual or 'screwdriver' inspection.

Quite apart from the potential inaccuracy of a visual inspection, it is first necessary to locate and expose the timber, with all the concomitant damage and cost of opening up and subsequent making good. The use of appropriate non-destructive techniques does not necessarily eliminate all such work, but does greatly reduce the costs and damage associated with opening up, and subsequent making good (Fidler 1980, 3).

LOCATION OF TIMBER

Thermography

Before any timbers can be assessed, it is of course necessary to locate them. In the great majority of historic buildings, some or all of the structural timbers are concealed behind coeval or later finishes. In some cases, such as plastered downstand floor beams, their location and approximate size is readily apparent, but more often they are more difficult to locate, whether they form part of external walls, internal walls or floors/ceilings. Concealed timbers inevitably tend to be more vulnerable to the agents of decay, which tend to prefer dark unventilated conditions. This is particularly the case in external walls,

Figure 1 A photograph of a rendered urban building of indeterminate age and construction on the corner plot of a block of buildings of varying dates and styles. See also Colour Plate 13.

Figure 2 The thermographic overlay of the building in Figure 1 identifies a number of interesting points. See also Colour Plate 14.

whether they be discrete elements such as window lintels, or form part of a coherent frame. The potential problems are exacerbated where inappropriate finishes such as cement renders and non-breathing paints have been used.

New developments in infrared thermography are beginning to make a significant contribution to building diagnostics and building archaeology. The range of applications for thermography within these fields is extensive, and most fall outside the present discussion, but the accurate location of structural timber and other features within external and internal walls and floors etc, is a proven application with demonstrable benefits.

All bodies above absolute zero (approximately -275°C) radiate energy in the form of electromagnetic waves. The intensity of this radiation within the infrared waveband is related to the thermal properties of the body, its surface temperature and emissivity (capacity to emit radiation). The various materials and components that comprise the body (ie timber, brick, stone etc of a building) absorb, transfer and radiate energy at different rates depending on their composition and inter-relationship. Heat will always flow from warmer bodies to cooler, so creating a thermal gradient; the greater the temperature difference, the steeper the gradient. In the case of buildings, the effects of these variations can be detected in the radiation of the energy from the various surfaces. The physical properties of infra-red radiation in terms of reflection, refraction and transmission are comparable to those of visible light, and can be focused through a lens, albeit of considerably more complex and expensive materials than the glass lens of a photographic camera.

Ultimately, the resolution and sensitivity (which translates to quality and quantity of information obtained) depends on the performance of an array of sensors within the camera, which are cooled to a very low temperature, typically 77K (-198°C). The highest quality systems, necessary for accurate and reliable building investigation, use a 256 x 256 staring indium antimonide (InSb) focal plane array, which gives the widest dynamic range and the highest sensitivity, allowing temperature differences as small as 0.025°C to be identified. This temperature

sensitivity is critical, as the thermal differences across the surfaces of a structure at ambient temperature can be very small. The greater the thermal sensitivity and image resolution of the camera, the greater the level of information that can be obtained, and the wider the range of conditions in which the technique can be used.

In the case of external elevations, no access equipment is normally required as the imaging can be carried out remotely from ground level. The thermal image can be viewed 'live' on an LCD screen attached to the camera (as with ordinary video cameras) and output to a monitor. It can also be recorded to digital (preferable) or analogue videotape, or stored as a still image on a PCMCIA card within the camera for subsequent computer analysis and processing.

Using image manipulation software, the thermal image can be rectified, scaled and overlain onto drawings or photographs, to provide an accurate and permanent record of the concealed structure. As the camera produces a live image, it is also possible to plot the positions of concealed elements directly onto the relevant surfaces. Alternatively, by using portable computer hardware, images can be printed immediately on site to guide the detailed condition assessment techniques described below.

External walls

With timber-framed walls that have been rendered over, the significant differences in the thermal properties of the frame and the infill (if present) between the frame members usually result in very clear delineation of the presence and exact location of the frame members. Frequently, individual timber elements, and even complete frames, have been identified within walls where no such elements had previously been suspected.

Figure 1 shows a rendered urban building of indeterminate age and construction on the corner plot of a block of buildings of varying dates and styles. Figure 2 shows the same photograph overlaid with a thermal image of the gable elevation. The framing pattern is very clearly defined and identifies a number of interesting points.

Figure 3 The rear elevation of a property, constructed of brick to ground floor and cement-based roughcast to first floor. The date of the building and materials of the first floor are indeterminate. See also Colour Plate 16.

- The original roof line has been raised by one panel height.
- The chimney stack in the corner is independent of the frame.
- The present window openings do not relate to the framing pattern.
- The ground floor has been rebuilt in brick, and is probably an underbuilt jetty.
- The lateral wall of the building is brick.

Quite apart from its value as an archaeological record, each of these points may well be of significance in assessing the building's structural performance and condition. The precision with which the various timber elements can be located allows the use of a micro-drilling system (described below) to assess them accurately without the need for the removal of any render. If significant decay is located, only relatively small areas of render in the relevant areas need be removed to effect the necessary repairs. This approach may be likened to the use of keyhole surgery in medicine. In addition, such a survey can be important in assessing the merits of any proposal to remove existing render from an elevation. Figure 3 shows a building with no indication whatsoever of the presence of timber framing. Figure 4 shows a thermographic image superimposed on the elevation, and readily identifies the timber frame (including down braces) and the positions of staircase and second floor. It also identifies

Figure 4 The same elevation as in Figure 3 with a thermographic image superimposed on the photograph. Immediately, and with no opening up or special access, the method of construction, the exact location of the individual elements and other useful information are clearly defined. See also Colour Plate 17.

Figure 5 A rendered building in Thaxted, Essex, with a strong shadow cast across the lower part of the elevation.

significant heat loss from the radiator below the first-floor window.

A common misunderstanding is that a large temperature difference or thermal gradient is needed between the building and its surroundings, or between the inside and outside of the building, for thermography to be effective, and also that solar gain will prevent useful information being obtained. This is not necessarily the case, as the use of the most sensitive cameras can eliminate, or at least minimise these problems. If there are absolutely no temperature differences across a wall, then no image can be obtained, but this is very rarely the case. The photographic and thermographic images shown in Figures 5 and 6 were taken in bright sunlight with a strong shadow cast across half the elevation. By careful configuration of the camera settings, the solar effects have been eliminated, and the concealed timber frame clearly identified.

There are no hard rules in terms of optimum weather conditions and thermal gradient. An elevation that is completely saturated by rain is unlikely to yield any relevant information, but monitoring the wall thermographically as it dries out can provide a great deal of information that might not be so readily obtained with the wall completely dry. The different rates of evaporation from varying surfaces and the effects of different substrates will affect the thermal image. These conditions can usually be artificially created if required. In some situations, and depending on the composition of the internal and external finishes, more information can be obtained from the internal face of an external wall. A tile-hung wall, for example, may be difficult to assess externally, but relatively straightforward to assess from within the building.

In addition to the actual location of the timbers, it is also possible to identify structural changes, for example the removal or insertion of windows that could indicate areas where there is increased risk of degradation, or where the structural integrity of the frame has been

Figure 7 *An internal wall of Kenwood House, London. There is no indication of the structure behind the fine decorative plaster. See also Colour Plate 18.*

Figure 6 *Careful configuration of the thermographic camera eliminates the effects of solar gain, and the structural frame of the building is clearly defined.*

significantly compromised. Referring back to the building shown in Figure 2, experience of assessing other timber-frames of this pattern would suggest that the mid-span of the original tie-beam is an area particularly vulnerable to decay, and that the lack of continuity in the corner posts may have significant structural implications.

In masonry buildings, bonding and other timbers, as well as other structural features, can be located where the wall has been plastered or rendered direct onto the masonry. If the walls have been lined so that there is an air gap between the finish and the masonry, it is not usually possible to obtain any significant information, although areas of relatively higher moisture content and other anomalies, which may indicate more serious problems, can be identified.

Internal walls

Even in masonry buildings it is common for a proportion of the internal walls to be framed in timber. While internal walls tend to be less prone to degradation than external walls, deathwatch beetle and other decay agents can cause substantial damage, particularly where internal walls abut chimney stacks or adjoin external walls. In addition, it can be important to establish the pattern of framing, and any changes that have taken place. For example, trussed walls are occasionally severely compromised by the insertion of a new doorway or other alteration. Figures 7 and 8 show photographic and thermal images of an internal wall at Kenwood House. The positions and pattern of structural timber shows very clearly behind the plain and ornate plaster. The thermographic image was obtained without any artificial heat sources.

The temperature differences between each side of an internal wall are usually smaller than across an external wall. Even so, the most sensitive cameras will usually be able to identify the significant structural elements within the wall, without the need for any introduced heat sources. When necessary, it is relatively easy to change the temperature on one side of the wall to create a thermal gradient. The temperature change does not need to be large; a rise of 2–4°C is all that is required. Where warming the entire room is not practical, low temperature radiant heaters or similar can be used, but must be treated with caution: there are occasions when too steep a thermal gradient will actually obscure the information required. For example, the identification of overpainted murals can only be achieved where the temperature differences are very small; the body heat generated by the camera operator will often be sufficient.

Floors and ceilings

The relatively low thermal conductivity of wood limits the location of structural timbers below floorboards, but

Figure 8 *The thermographic image superimposed on the photograph clearly shows the structure, and its exact location. See also Colour Plate 19.*

Figure 9 The thermal image of this church ceiling identifies the secondary timbers concealed behind the plaster. More significantly, a damp patch (arrowed) indicates a leak in the roof that could not be accessed from above the ceiling. Had this not been identified, the persistent wetting would have encouraged fungal and insect attack. See also Colour Plate 15.

thermography has been used successfully to locate pipes and overloaded or faulty electric cables below floor-boards, and even to locate rodent nesting sites.

The location of floor/ceiling timbers from below, where concealed by plaster, is relatively straightforward, and the pattern and positions of principal and secondary members can normally be readily located. This is an important application, not just for the assessment of load-bearing elements of a floor, but also for the identification of timber grounds behind ornate plaster, and for locating areas where plaster ceilings are sagging away from the supporting structure.

The identification of complete or fragmented timber frames and elements behind render and plaster is a very useful and relatively straightforward application. More generally, the ability of thermography to identify high moisture levels, failures in render, cracks in masonry etc within the structural envelope of all building types, greatly helps to identify those areas most at risk from fungal and insect attack. For example, failures in external render can lead to significantly higher localized moisture levels. If these wetter areas correspond with positions of internal floor beams built into the wall, this clearly indicates areas where particularly thorough assessment of the timber should be carried out. Similarly, the effective-ness of any measures adopted to reduce the moisture levels within the building envelope, perhaps as part of a

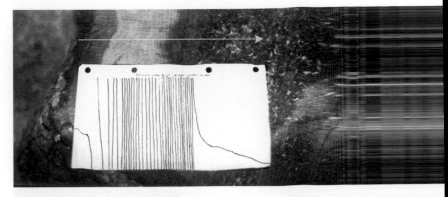

Figure 11 The paper trace from a microdrill laid against the timber tested. The variations in density can be clearly identified.

strategy to control deathwatch beetle activity, can be monitored over time. Figure 9 shows the ceiling of church: the secondary rafters/joists can be seen, an water penetration can be identified (arrowed) which would not have been identified by visual inspection unt it had caused significant degradation. Where buildings ar known to be at risk from deathwatch beetle infestation the use of thermography to identify and monitor env ronmental factors such as water penetration is a valuab tool.

Limitations

Difficulties in obtaining information thermographical arise when there is a thermal break in the wall constru tion, and/or the external surface of the wall is of relative very low conductivity. For example, if a brick skin h been built outside the line of the timber-frame (between the frame and the thermographic camera) wit relatively little connection between, it is unlikely th there will be any indication of the timber behind. How ever, this is not the case where timber elements such lintels are actually built into the wall, where their pre ence will have an effect on the overall thermal mass of th wall. This ability to identify lintels and bonding timbers bu into masonry walls is particularly valuable, as they are ofte very vulnerable to fungal and deathwatch beetle attac which in turn can lead to severe structural instability.

It can be difficult to predict which combination weather conditions and other factors will produce th most useful results on any given day for a particul building type. Generally, periods of greater temperatu variation are more likely to prove successful. As discusse above, a fully saturated wall, or wall surface, will yiel very little information while it remains saturated, an high winds will also tend to reduce the thermal diffe ences on the surface. Internally, suspended ceilings an dry linings to walls will obscure any structure behind, will thermal, and some types of sound insulation. A these limitations can be minimised by the use of the mo sensitive cameras.

In terms of interpretation, it must always be borne i mind that, while the presence of identifiable elements structure is clear evidence of their existence, the appare absence of identifiable elements does not necessaril

Figure 10 The Sibert DDD200 microdrill.

Figure 7 An internal wall of Kenwood House, London. There is no indication of the structure behind the fine decorative plaster. See also Colour Plate 18.

Figure 6 Careful configuration of the thermographic camera eliminates the effects of solar gain, and the structural frame of the building is clearly defined.

significantly compromised. Referring back to the building shown in Figure 2, experience of assessing other timber-frames of this pattern would suggest that the midspan of the original tie-beam is an area particularly vulnerable to decay, and that the lack of continuity in the corner posts may have significant structural implications.

In masonry buildings, bonding and other timbers, as well as other structural features, can be located where the wall has been plastered or rendered direct onto the masonry. If the walls have been lined so that there is an air gap between the finish and the masonry, it is not usually possible to obtain any significant information, although areas of relatively higher moisture content and other anomalies, which may indicate more serious problems, can be identified.

Internal walls

Even in masonry buildings it is common for a proportion of the internal walls to be framed in timber. While internal walls tend to be less prone to degradation than external walls, deathwatch beetle and other decay agents can cause substantial damage, particularly where internal walls abut chimney stacks or adjoin external walls. In addition, it can be important to establish the pattern of framing, and any changes that have taken place. For example, trussed walls are occasionally severely compromised by the insertion of a new doorway or other alteration. Figures 7 and 8 show photographic and thermal images of an internal wall at Kenwood House. The positions and pattern of structural timber shows very clearly behind the plain and ornate plaster. The thermographic image was obtained without any artificial heat sources.

The temperature differences between each side of an internal wall are usually smaller than across an external wall. Even so, the most sensitive cameras will usually be

able to identify the significant structural elements within the wall, without the need for any introduced heat sources. When necessary, it is relatively easy to change the temperature on one side of the wall to create a thermal gradient. The temperature change does not need to be large; a rise of 2–4°C is all that is required. Where warming the entire room is not practical, low temperature radiant heaters or similar can be used, but must be treated with caution: there are occasions when too steep a thermal gradient will actually obscure the information required. For example, the identification of overpainted murals can only be achieved where the temperature differences are very small; the body heat generated by the camera operator will often be sufficient.

Floors and ceilings

The relatively low thermal conductivity of wood limits the location of structural timbers below floorboards, but

Figure 8 The thermographic image superimposed on the photograph clearly shows the structure, and its exact location. See also Colour Plate 19.

Figure 9 The thermal image of this church ceiling identifies the secondary timbers concealed behind the plaster. More significantly, a damp patch (arrowed) indicates a leak in the roof that could not be accessed from above the ceiling. Had this not been identified, the persistent wetting would have encouraged fungal and insect attack. See also Colour Plate 15.

thermography has been used successfully to locate pipes and overloaded or faulty electric cables below floorboards, and even to locate rodent nesting sites.

The location of floor/ceiling timbers from below, where concealed by plaster, is relatively straightforward, and the pattern and positions of principal and secondary members can normally be readily located. This is an important application, not just for the assessment of load-bearing elements of a floor, but also for the identification of timber grounds behind ornate plaster, and for locating areas where plaster ceilings are sagging away from the supporting structure.

The identification of complete or fragmented timber frames and elements behind render and plaster is a very useful and relatively straightforward application. More generally, the ability of thermography to identify high moisture levels, failures in render, cracks in masonry etc within the structural envelope of all building types, greatly helps to identify those areas most at risk from fungal and insect attack. For example, failures in external render can lead to significantly higher localized moisture levels. If these wetter areas correspond with positions of internal floor beams built into the wall, this clearly indicates areas where particularly thorough assessment of the timber should be carried out. Similarly, the effectiveness of any measures adopted to reduce the moisture levels within the building envelope, perhaps as part of a

Figure 10 The Sibert DDD200 microdrill.

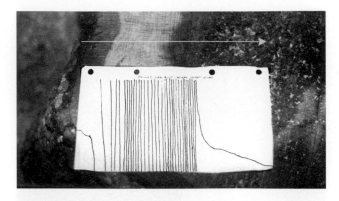

Figure 11 The paper trace from a microdrill laid against the timber tested. The variations in density can be clearly identified.

strategy to control deathwatch beetle activity, can be monitored over time. Figure 9 shows the ceiling of a church: the secondary rafters/joists can be seen, and water penetration can be identified (arrowed) which would not have been identified by visual inspection until it had caused significant degradation. Where buildings are known to be at risk from deathwatch beetle infestation, the use of thermography to identify and monitor environmental factors such as water penetration is a valuable tool.

Limitations

Difficulties in obtaining information thermographically arise when there is a thermal break in the wall construction, and/or the external surface of the wall is of relatively very low conductivity. For example, if a brick skin has been built outside the line of the timber-frame (ie between the frame and the thermographic camera) with relatively little connection between, it is unlikely that there will be any indication of the timber behind. However, this is not the case where timber elements such as lintels are actually built into the wall, where their presence will have an effect on the overall thermal mass of the wall. This ability to identify lintels and bonding timbers built into masonry walls is particularly valuable, as they are often very vulnerable to fungal and deathwatch beetle attack, which in turn can lead to severe structural instability.

It can be difficult to predict which combination of weather conditions and other factors will produce the most useful results on any given day for a particular building type. Generally, periods of greater temperature variation are more likely to prove successful. As discussed above, a fully saturated wall, or wall surface, will yield very little information while it remains saturated, and high winds will also tend to reduce the thermal differences on the surface. Internally, suspended ceilings and dry linings to walls will obscure any structure behind, as will thermal, and some types of sound insulation. All these limitations can be minimised by the use of the most sensitive cameras.

In terms of interpretation, it must always be borne in mind that, while the presence of identifiable elements of structure is clear evidence of their existence, the apparent absence of identifiable elements does not necessarily

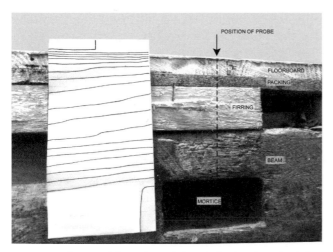

Figure 12 The microdrill can be used through floorboards to check the construction and condition of underlying structure.

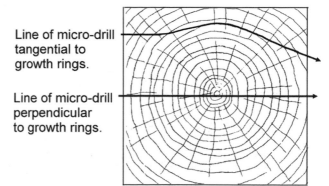

The effect of growth ring orientation on micro-drill probes.

Figure 13 The flexible probe of the microdrill will sometimes track round growth rings, particularly in fast-grown softwoods.

mean that they are not there. However, thermography has been used successfully to demonstrate that a building thought to be timber-framed was in fact constructed of flint and brick, as there was sufficient clear evidence of these materials behind the render.

Once the timbers have been located, a detailed assessment of their condition can be carried out.

ASSESSMENT OF TIMBERS

Micro-drilling

The Sibert microdrilling equipment uses the same basic principle as any drill or probe used to assess timber, in that it detects variations in resistance to penetration caused by changes in the quality of the timber. What makes the Sibert systems so much better is the very large amount of precise information obtained, the speed with which it can be obtained and the very small amount of damage caused. The microdrill was originally developed for locating and assessing a particular fungus (*Heterobasidion annosum*) in commercially grown conifers. Subsequently it has proved itself capable of providing valuable information in a wide range of timber assessment applications.

A new digital version (the DmP) has been developed from the original model (the DDD200). Both work on exactly the same principle: the key differences lie in the methods of presentation and quantity of data retrieved, and are discussed below.

In essence the micro-drill consists of a 1 mm diameter probe rotating at 7000 rpm which is pushed at a controlled and continuously monitored pressure into the timber under test (Fig 10). With the DDD200 the rate of penetration is recorded by pen onto a paper sheet on a slowly revolving drum: this gives an immediate and permanent record of the changes in condition throughout the depth of timber tested, from which a great deal of valuable information can be obtained (Fig 11). The density of the timber can be assessed, and sudden or significant variations in the trace indicate the presence of decay or other changes within the timber. The depth at which these changes occur can be measured directly from

the trace (Fig 12). If required, an accurate cross-section of the timber, showing the position and extent of any degradation, can be derived by carrying out a series of drillings at different positions and in various directions.

The tip of the probe is profiled but not sharp, and it does not cut into the wood as a conventional drill does: as a result the rate of penetration is not affected by any blunting that might occur. The probe itself is flexible, and supported along its length by sliding bearings and the timber itself. In spite of the probe's flexibility, the hole drilled into the timber generally remains straight, particularly in hardwoods. However, in softwoods there is sometimes a tendency for the probe to follow the relatively less dense early wood of an annual growth ring, giving misleading results that may be interpreted as decay by the inexperienced. This phenomenon can be minimized if the probe travels at right angles through the growth rings, but this can be difficult to maintain towards the heart, where the rings are relatively small (Fig 13). With experience, it is usually straightforward to identify when it occurs.

With the DDD200, the maximum depth of penetration into the timber is normally set at 200 mm (8 in), which is adequate for most situations. If necessary the penetration depth can be increased in 200 mm (8 in) increments up to 600 mm (23.5 in), but full assessment at this depth is time-consuming. Typically, an assessment up to 200 mm (8 in) penetration takes less than 30 seconds. Although not strictly non-destructive, the hole left by the probe is less than 1 mm in diameter, and is virtually indistinguishable from a furniture beetle flight hole. One of the great advantages of the microdrill is that testing can be carried out through finishes such as floorboards, panelling and plaster (see Fig 14), with negligible disturbance or damage to those finishes. A typical example where this facility is of particular value is the assessment of lintels concealed behind plaster or timber linings.

If the probe enters a cavity within the timber, a pressure switch automatically stops the probe rotating until the tip reaches the other side of the cavity. In this way, the size of any cavity (or mortice etc) can be

Figure 14 Timbers behind or within decorative plaster and other finishes can be tested with negligible damage to the finish.

accurately measured (see Fig 12). Similarly, the probe stops rotating when the tip emerges from the timber, so the actual dimensions of the timber can also be accurately measured. This is very useful where timbers are completely or partially concealed. Because the probe is so fine there is a risk of it buckling if it remains unsupported for more than 60–70 mm (*c* 2–2½ in). Also, if there is a void greater than 60–70mm (*c* 2–2½ in) between the coverings and the timber (as often occurs, for example between window linings and lintels), it may be necessary to drill a 2.5 mm pilot hole through the covering, and use an adapted probe that incorporates a 2 mm internal diameter tube to support the probe in the void between the covering and the timber. Where such pilot holes need to

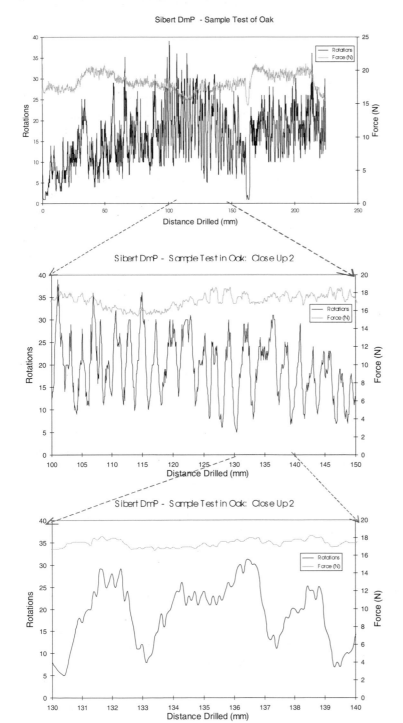

Figure 15 A chart from a digital probe into oak to a depth of 225 mm (9 in) The chart shows the number of rotations of the probe per 0.1 mm of penetration and the force applied which is kept relatively constant by the operator. Minor variations relate directly to changes in the actual timber and are recorded on the chart. There is a fault (probably a shake) between 161 mm and 164 mm.

Figure 16 A detail of the chart in Figure 10 between 100 mm (4 in) and 150 mm (c 6 in) in more detail. Each annual growth ring is very clearly defined. With further research and development, this could allow accurate dendrochronological dating without the need for destructive core sampling. Many other factors such as methods of woodland management may also be better understood from the data obtained. Research is continuing.

Figure 17 A closer detail of the chart in Figure 10 between 130 mm and 140 mm. The major peaks and troughs clearly define each annual growth ring, and their thickness can be accurately measured. The significance and relevance of the minor peaks and troughs is not yet fully understood, but may well provide sufficient additional data to produce reliable measurements of other relevant criteria such as Young's Modulus and dendrochronological dating.

be used, it may be more accurate to describe the technique as micro-destructive rather than non-destructive, but each pilot hole is only 2.5 mm diameter, and it remains vastly less damaging than the macro-destructive alternative of opening up for visual inspection.

The probe is capable of penetrating relatively soft lime plasters without a pre-drilled pilot hole, but the abrasive nature of the plaster will tend to change the profile of the tip relatively quickly. In an assessment carried out on a very ornate seventeenth-century plaster ceiling, the three coats of plaster could be clearly differentiated and their thicknesses measured (Fig 14), but this information is not normally required, and a 1.5 mm pilot hole should be drilled (no support tube is required). After the assessment, the pilot holes can be easily and quickly filled with lime putty or other suitable material if required.

A recent development, the Sibert DmP, works on much the same principle, but records the data digitally, rather than as a trace on paper (Fig 15). This eliminates the 200 mm (8 in) depth restriction created by the paper trace, allowing penetration to any depth up to 1 m (3 ft), without changing the probe. The digital record can be displayed graphically using standard spreadsheet software. The DmP contains a microprocessor that collects the data as the drilling takes place and transfers it to a portable computer via an RS232 cable. At every 0.1 mm of penetration, the distance drilled to that point, the number of revolutions taken to progress the 0.1 mm and the force applied at that point, are recorded as the probe travels through the timber.

The most significant difference between the digital version and the original micro-drill lies in the quantity of data obtained. Figure 15 shows a chart derived from the data obtained from a Sibert DmP drilling into a sample of oak. Figures 16 and 17 show the details of the same chart at different scales. The differences between early and late growth within each annual ring and the variations between annual rings are clearly defined, which allows a far more accurate assessment of the quality of the timber. Fast-grown softwoods are significantly weaker and less durable than slow-grown samples of the same species, but the reverse is true of temperate hardwoods such as oak. In many situations it is useful to be able to measure the rate of growth of the timber, and so derive a reasonable measure of its quality.

The smaller peaks and troughs within each annual ring may represent the natural variations within the cell structure of the timber, but may also be affected to some extent by other factors such as seasonal variations in weather conditions. With the level of detail obtainable using a DmP, it might well be possible to identify attacks by fungi such as *Fistulina hepatica*, the presence of which is thought to have an effect on the incidence of death-watch beetle attack, and is normally very difficult to detect. In addition, the microscopic structural changes that seem to have taken place in medieval oak may be better understood, and their significance in relation to deathwatch beetle attack and other decay mechanisms more clearly resolved. These and many other areas require further research.

The data obtained should also provide a basis for accurate dendrochronological dating. This area is the subject of a major research project, currently being undertaken by Demaus Building Diagnostics, the initial results of which should be available by the end of 2001. If this research proves that accurate and reliable dating can be achieved using the DmP, it would allow far more comprehensive dating surveys to take place.

For straightforward condition assessment of oak, the digital version may provide an unnecessary volume of data, but spreadsheet programmes can filter the data obtained to give the level of detail required for any particular application.

Both the DDD200 and the DmP are fully portable, with their own battery packs. The DmP is the more compact, but does need a computer or data logger attached.

Ultrasound

Ultrasound is one of the most widespread non-destructive techniques for materials testing, and is routinely used in many industrial, scientific and medical applications. Within the field of building conservation it has applications in the assessment of stone, concrete, metals and ceramic materials, as well as timber. The equipment varies considerably depending on requirements for the material tested, but essentially all consist of a transmitter that emits pulses of ultrasound, a receiver that picks up the signals and equipment that measures the time taken for the ultrasonic pulse to travel through the material between them (the transit time), or echo back from boundary changes within the material. In some applications, this data can then be processed to produce visual images (perhaps the best-known example being foetal examination in antenatal clinics).

While it would be highly desirable to obtain a visual image of the internal condition of timber, the particular physical properties of wood as a material, as well as the restrictions imposed by site investigation and conservation issues, render this impracticable.

Timber is one of the more difficult materials to assess ultrasonically for a number of reasons. In addition to the considerable natural variations caused by factors such as growing conditions, timber is orthotropic, ie the ultrasonic properties vary in the longitudinal, radial and tangential directions.

With homogenous and generally isotropic materials such as metal, relatively high-frequency (200 kHz–1000 kHz+) ultrasound can be used as the ultrasonic 'beam' is not attenuated or scattered, therefore internal defects can be detected and accurately located by the echoes reflected back in the direction of the incident beam, and which are picked up by the receiving transducer. The time taken for the pulse to travel from the transmitter to the defect and back to the receiver enables the depth at which the defect occurs to be accurately calculated. This method cannot be applied to heterogeneous and relatively unpredictable materials such as timber, because the echoes generated at the numerous boundaries of the different phases scatter

Figure 18 Ultrasound being used to test the integrity of a large cruck blade.

the pulse energy in all directions. However, by using relatively low frequency ultrasound (generally 20 kHz–100 kHz) and placing the transducer heads on opposite faces of the timber, it is possible to measure the transit times, and thus quickly obtain a reliable indication of the condition of the timber (Fig 18). Different species of timber have characteristic pulse velocities, and the transit times in a fault-free sample of each species are directly proportional to the thickness of the sample (ie the distance between the transducer heads) (Fig 19). However, because of the natural variations within the timber, changes in transit time within 15% either side of the characteristic pulse velocity are usually ignored: also, the pulse is slightly faster if the pulse travels perpendicular to the growth rings (Fig 20). It is the significant deviations from the transit time (typically more than 25%) that identify areas of decay or other anomalies within the timber. Considerable experience is required in interpreting the data and differentiating between changes in transit time caused by decay and what may be acceptable anomalies (such as splits) in the timber. Ultrasound is completely non-destructive, leaving no stain or other damage, and can be used on fragile decorative finishes

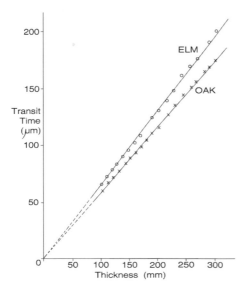

Figure 19 In any given timber, the characteristic pulse velocity remains constant, so the transit time is directly proportional to thickness.

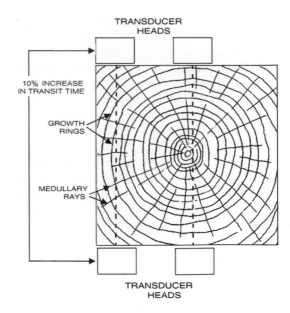

Figure 20 The anisotropy of wood creates variations in the transit time related to grain orientation and growth rings etc.

such as paint and lacquer where other techniques might cause problems.

For general assessment work on oak, the microdrilling systems tend to provide more comprehensive information, but ultrasound can provide a very useful check where microdrilling is inconclusive, or where even the very small hole created by the microdrill is unacceptable. In softwood structures such as nineteenth-century roofs, ultrasound is a very quick and reliable assessment method. It is also useful in the assessment of glued laminates.

There are a number of specific applications where two or more assessment techniques used together are needed to obtain all the relevant information. An example is the assessment of joints in timber frames. Most joints, particularly mortice and tenon joints, tend to be vulnerable to deathwatch beetle attack, and also, on external walls, to fungal decay. Frequently, degradation will only affect one element of the joint, with no visible indication of whether the pegs and/or the tenon have failed. Where the joint remains in compression at all times, this is not necessarily a significant problem, but where there is any chance of the joint acting in tension under certain load conditions, even temporarily, the condition of the pegs and tenon becomes critical: bell frames and arch brace roof trusses are typical examples. The micro-drill is an ideal tool for assessing the joint elements, but cannot be used to check the condition of the pegs, particularly when draw-bored. Ultrasound, on the other hand, is very effective at identifying fractured or severely degraded pegs without any disturbance to the joint, but cannot check the tenon housed within the mortice.

SUMMARY

Thermography

The use of the most sensitive thermographic cameras can provide an astonishing range of information on the archaeology, structure and performance of historic buildings. Particular applications relevant to the assessment of timber include the location of frames and individual timber elements within walls and floors, and the identification of damp and failures in the building envelope likely to encourage decay mechanisms.

Microdrilling

The microdrill is probably the most useful single tool currently available for the assessment of timber. With further research, the new digital microdrill should provide additional valuable information on strength, density, decay mechanisms and dating.

Ultrasound

Ultrasound provides an effective assessment of faults within timber, which is faster than microdrilling, but without the precision. It is particularly effective on softwood structures and has specialist applications that include the assessment of pegs, glue-laminated beams and delicate and/or painted timber.

In many situations the combination of different techniques provides the widest range of useful information.

CONCLUSIONS

The methods described above are analogous to the various diagnostic techniques used in medicine as a prerequisite to any decision on treatment or surgery. The benefits of accurate diagnosis are as demonstrable in building repair and conservation as they are in medicine. They allow accurate location and assessment of the condition of concealed and expressed structural timber and decorative timber. The use of the correct non-destructive assessment methods almost invariably results in significant reduction in damage, and loss of original and/or historically important fabric. In most cases, accurate investigation will also greatly diminish unforeseen problems arising once a contractor is on site, which so often result in delays and increased costs.

It should be stressed that the best time to employ the sort of techniques described above is at the start of a project, before any decisions about the extent and method of repair are taken, and preferably before any stripping out/opening up contract has been let. In certain cases it may be necessary to return to some areas during the project as access becomes available, but a large proportion of the information, which may have significant influence on the decision-making process, can, and should, be obtained at an early stage. Quite apart from minimising material damage, a thorough non-destructive survey at the start of a proposed project will usually result in cost savings far in excess of the actual cost of the survey.

BIBLIOGRAPHY

Fidler J, 1980 Non-destructive surveying techniques for the analysis of historic buildings, in *Transactions of the Society for Studies in the Conservation of Historic Buildings*, 3–10.

EQUIPMENT

Thermographic imaging camera
Amberaytheon c/o LOT Oriel, 1 Mole Business Part, Leatherhead KT22 7AU; +44 (0)1372 378833.

Sibert Microdrill
Sibert Technology Ltd, 2A Merrow Business Centre, Merrow Lane, Guildford GU4 7WA; +44 (0)1483 440724 www.sibtech.com

PUNDIT Tester
CNS Electronics, 61–3 Holmes Road, Kentish Town, London NW5 3AL.

AUTHOR BIOGRAPHY

Robert Demaus (BEng, PG Dipl Cons, IHBC) specializes in the investigation and assessment of historic buildings and other structures, using a range of non-invasive, non-destructive and micro-destructive techniques. He regularly lectures to conservation practitioners and students on non-destructive investigation methods, and is involved in researching and developing new techniques.

Monitoring conditions and treatments of buildings
Case studies from The Netherlands

PETRA ESSER ★ AND JAN DE JONG

TNO Building and Construction Research, PO Box 49, 2600 AA Delft, The Netherlands;
Tel: + 31 15 284 2000; Fax: + 31 15 284 3990.

Abstract

The efficacy of heat treatments for eradication were monitored *in situ* in churches with deathwatch beetle infections.

Key words

Deathwatch beetle, monitoring, heat treatment.

INTRODUCTION

Three case studies of monitoring and treatments of buildings in The Netherlands are highlighted to show the wide spectrum of problems that may occur and the different available techniques for monitoring and treatment of timber decay through beetle infestation. A stepwise approach for development of renovation and treatment schedules of deathwatch beetle infections in buildings is proposed.

Several case studies were carried out to improve the consistency and the quality of inspection and monitoring of deathwatch beetle and other infections of wood in historic buildings.

Monitoring of moisture in wood structures and the surrounding building components play an important role in preparation, execution and follow-up of treatment of deathwatch beetle infestation. Results of a year of monitoring of two churches in The Netherlands revealed that temporary heating and introduced moisture (from wet clothes) during meetings does not have an effect on the moisture content of the wood (Castenmiller & de Jong 1997b). If the buildings and the oak are dry, wood decay will not proceed any further. Strength assessment using DDD-drills should be interpreted in relation to the angle of the drill-axis and the pattern of the year rings within the oak beams.

Two types of remedial treatment proved to be successful: the Wijhe heating method (Casternmiller & de Jong 1997a), using wet heat up to 55°C, or injection of deltamethrin in micro-emulsion combined with surface treatment as prevention to re-infestation (Esser & de Jong 1996). This research is continuing, outside the Woodcare project.

TRIPPENHUIS (1660), AMSTERDAM

In 1996 TNO advised on an assessment report of the Trippenhuis, a famous historic building at Amsterdam (Castenmiller & de Jong 1996). It was used as a case study

★ Author for correspondence

for the Woodcare project. This structure had just been renovated (1989–91), when some construction failures appeared. An inspection of the oak construction with the DDD-drilling bore resulted in an alarming inspection report, in which the strength and safety of the building's construction was questioned. It was thought that the construction contained active fungal decay and deathwatch beetle attack.

TNO introduced a three-step research approach:

- detection of the cause of decay, by monitoring of moisture of a selection of wooden beams, joining walls and visual inspections
- assessment of the type and extent of decay by visual inspection, sound detection and wood sample analysis
- assessment of the residual strength of the wood structures using a DDD-drilling bore.

The conclusions from this case study (Mulder 1997) were:

- The moisture content of the wooden beams was generally below 20% and could be considered dry. The walls did not show any sign of moisture traps.
- The fungal decay and deathwatch beetle decay were both old and not active, and treatment was not needed at this time.
- The residual strength of beams was sufficient, with one exception in one room (Bilderdijk-room), but many beams had large cracks.
- It was shown that the former alarming inspection report was mainly based on misinterpretations of DDD-drilling graphs. Large cracks and tangential orientation of year rings may lead to misleading DDD-graphs. Therefore it is important to drill at more than one side of a beam to obtain a complete picture. In case of doubt it is advised to use two or more persons for drilling assessments.

At present the techniques used for assessment of residual strength still need further improvement, and research at TNO on this subject is continuing.

FRYSIAN CHURCHES: MOISTURE MONITORING

A building's microclimate is sometimes suspected to be the cause of high moisture content in the wood, for

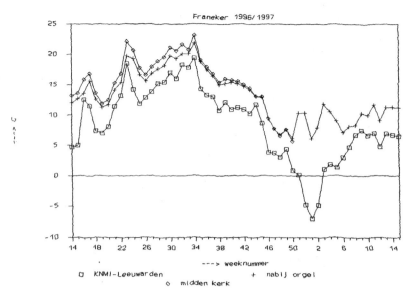

Figure 1 *Franeker Church, weekly average temperature of the air inside the church.*

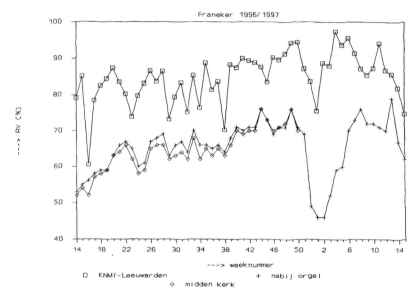

Figure 2 *Franeker Church, weekly average relative humidity (expressed as %).*

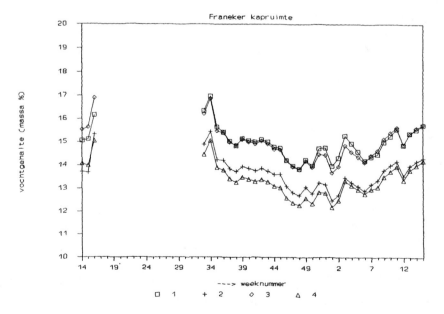

Figure 3 *Franeker Church, weekly average moisture content of four selected areas in the oak roof structure*

example in churches. Moisture may be introduced by visitors to the church, by both respiration and damp clothes, with an important effect on the building's climate and thus on the wood's moisture content. The churches of Berlikum and Franeker in Frysland, both with active deathwatch beetle infestation, were monitored for a year (1996-97). Temperature, relative humidity and the moisture content of selected wooden beams were measured (Figs 1, 2 and 3).

The results of the measurements show that church visitors do not significantly influence the building climate. The relative humidity of the climate in the church is determined mainly by the outside climate and the temperature inside the church. The relative humidity of both churches varies between 55% and 75%. If a high moisture content of the wood is found, there must be other moisture sources (eg leakages, moisture in walls). Heating of churches during frost periods will lead to a relatively fast decrease of the relative humidity. This may lead to damage to organs and other works of art in the church.

WIJHE HEATING METHOD AT SLOTEN

Active deathwatch beetle larvae in oak structures may be treated by wet heat (minimum 24 hours at 55°C). This method is known in The Netherlands as the Wijhe heating method. There has been considerable doubt about the efficacy of heating treatments, as to whether the heat would reach the inside of the large oak beams, where deathwatch beetle is often most active. The church at Sloten was chosen as a case study to measure the efficacy by assessment of the activity of larvae in the wood before and after the treatment. This was performed using the new acoustic detection method (see van Staalduinen et al, this volume, Fig 4). The heat treatment was carried out by De Slegte at Wijhe, The Netherlands (Figs 5 and 6). The organ was wrapped in plastic to prevent any damage.

The result of the acoustic detection assessment at three selected locations in the church indicate that the treatment was successful. It was noted that the low relative humidity may lead to cracks in thin wooden panels. A

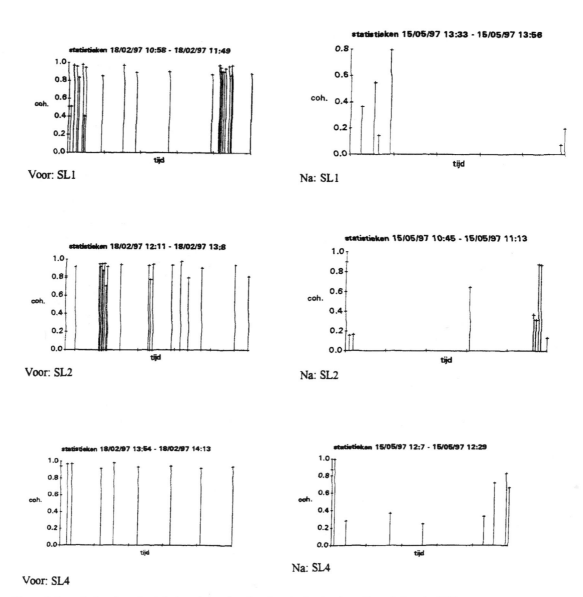

Figure 4 Acoustic detection: Statistical results at three locations in the church at Sloten before the Wijhe heating treatment (three graphs on the left) and after the treatment (three graphs on the right).

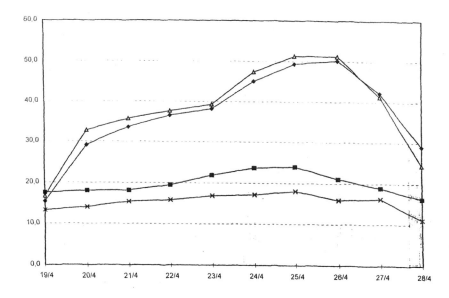

Figure 5 Temperature measured in the church at Sloten during the Wijhe heating treatment. Lines from top to bottom: temperature of the wood, the organ, the air in the church and the air within the organ.

Figure 6 Heat treatment sensor at the church at Sloten, The Netherlands.

darkening of the colour of the oak was noted, which may have been due to reactions of extractives.

For future use the physical building conditions should be checked to ascertain whether they are appropriate for this heating method. Furthermore, the heating method does not give protection against re-infestation by death-watch beetle. It is therefore advised to combine this heating method with a surface treatment (spraying) with a water-soluble insecticide (eg deltamethrin). A yearly visual inspection and acoustic detection is still recommended.

CONCLUSIONS

A three-step plan is advised to set up plans for treatment and renovation:

- Find the cause
- Assess type and extent of wood decay
- Assess residual strength of structures.

The Wijhe heating method has been effective in treatment of deathwatch beetle larvae. Injection of deltamethrin in micro-emulsion solution has been effective in remedial treatment. Surface treatment is applied also after injection, against re-infestation.

BIBLIOGRAPHY

Castenmiller C J J and de Jong J D, 1996 *Research on the Condition of the Wooden Structure of the Trippenhuis and Klovernierburgwal at Amsterdam*, TNO report 96-CHT-R1571 (in Dutch), Dutch State Building Service.

Castenmiller C J J and de Jong J D, 1997a *Research on the Efficacy of the Heat Treatment according to Wijhe, in a Church at Sloten,* TNO Report 97–CHT–R0177 (in Dutch), Dutch Foundation for Research into the Deathwatch Beetle.

Castenmiller C J J and de Jong J D, 1997b *Measurements on the Moisture Conditions of two Frysian Churches,* TNO report 97–CHT–R0179 (in Dutch), Dutch Foundation for Research into the Deathwatch Beetle.

Esser P M and de Jong J D, 1996 *Research on the Efficacy of Water-Dispersable Products (Micro-Emulsions) in Eradication of Insect Attack of Wooden Structures,* TNO Report 95–CHT–R1311 (in Dutch), contracted by the Ministry of Housing, Physical Planning and Environment in the VOC Research Programme (KWS 2000).

Mulder, ir H J, 1997 *Final Report of the Problems with the Wooden Structure of the Trippenhuis* (in Dutch), summary report by the Dutch State Building Services.

AUTHOR BIOGRAPHIES

Petra Esser is leader of the Wood and the Environment working group in the Centre for Timber Research at TNO Building and Construction Research in The Netherlands. She has an MSc degree in Environmental Biology from the University in Leyden and over 10 years of experience in biological and environmental research. Research topics range from degradation processes of wood, effectiveness of anti-sapstain treatments and impregnation of wood, measuring of emissions from preservative treated wood, to performance of complete life cycle assessments of wooden products. Dr Esser was the TNO coordinator of the Woodcare project and prepared laboratory and field wood samples for chemical analysis.

Jan de Jong is leader of the working group Wood in Buildings in the Centre for Timber Research of TNO Building and Construction Research, The Netherlands. His educational background is in building technology. He has 15–20 years of experience of working at TNO on damage assessment in wooden structures and wooden building interiors. He also advises on repairs and treatment of wood and is involved in testing the durability of products and product development. Projects are performed for builders, industrial clients, government (national and the EU) and building co-operatives. His role in the Woodcare project was monitoring conditions in buildings, detection of deathwatch beetle larvae in laboratory and field conditions (with Piet van Staalduinen) and testing heat treatments.